T0292008

Hydropower

The Power Generation Series

Paul Breeze—Coal-Fired Generation, ISBN 13: 9780128040065
Paul Breeze—Gas-Turbine Fired Generation, ISBN 13: 9780128040058
Paul Breeze—Solar Power Generation, ISBN 13: 9780128040041
Paul Breeze—Wind Power Generation, ISBN 13: 9780128040386
Paul Breeze—Fuel Cells, ISBN 13: 9780081010396
Paul Breeze—Energy from Waste, ISBN 13: 9780081010426
Paul Breeze—Nuclear Power, ISBN 13: 9780081010433
Paul Breeze—Electricity Generation and the Environment, ISBN 13: 9780081010440

Hydropower

Paul Breeze

ACADEMIC PRESS

An imprint of Elsevier

Academic Press is an imprint of Elsevier
125 London Wall, London EC2Y 5AS, United Kingdom
525 B Street, Suite 1800, San Diego, CA 92101-4495, United States
50 Hampshire Street, 5th Floor, Cambridge, MA 02139, United States
The Boulevard, Langford Lane, Kidlington, Oxford OX5 1GB, United Kingdom

Notices
Knowledge and best practice in this field are constantly changing. As new research and experience
broaden our understanding, changes in research methods, professional practices, or medical treatment
may become necessary.

Practitioners and researchers must always rely on their own experience and knowledge in evaluating
and using any information, methods, compounds, or experiments described herein. In using such
information or methods they should be mindful of their own safety and the safety of others, including
parties for whom they have a professional responsibility.

To the fullest extent of the law, neither the Publisher nor the authors, contributors, or editors, assume
any liability for any injury and/or damage to persons or property as a matter of products liability,
negligence or otherwise, or from any use or operation of any methods, products, instructions, or ideas
contained in the material herein.

British Library Cataloguing-in-Publication Data
A catalogue record for this book is available from the British Library

Library of Congress Cataloging-in-Publication Data
A catalog record for this book is available from the Library of Congress

ISBN: 978-0-12-812906-7

For Information on all Academic Press publications
visit our website at https://www.elsevier.com/books-and-journals

Working together
to grow libraries in
developing countries

www.elsevier.com • www.bookaid.org

Publisher: Joe Hayton
Acquisition Editor: Lisa Reading
Editorial Project Manager: Mariana Kuhl
Production Project Manager: Vijayaraj Purushothaman
Cover Designer: MPS

Typeset by MPS Limited, Chennai, India

CONTENTS

CHAPTER *1*

An Introduction to Hydropower

Hydropower was probably the first renewable energy resource in the world to be exploited and one of mankind's first sources of mechanical power. There are indications that the use of water power may have been known in Mesopotamia as early as 4000 BC although interpretation of the evidence is difficult. Clearer evidence can be found in the first millennium BC. The earliest known literary reference is found in a Greek poem of 85 BC and there are occurrences in Roman texts too. Egyptian papyri from the second and third centuries BC also indicate the use of water wheels. Simple wheels used to drive mills and grind grain were known in China during the first century AD and by the beginning of the second millennium the technology was used widely throughout Asia and Europe. The coupling of water wheels with generators to provide electric power was initiated at the end of the 19th century in Europe and the United States.

Early mills and water wheels were relatively simple devices that used wooden paddles. Iron was introduced in the 18th century during the industrial revolution in England. This innovation led to the development during the 19th century of many of the more advanced and efficient turbines now in use in modern hydropower stations.

Hydropower capacity grew strongly during the 20th century and until late in that century it was the only significant renewable source of electrical power. According to the International Hydropower Association (IHA) the total global installed capacity of hydropower plants stood at 1246 GW at the end of 2016, including pumped storage hydropower.[1] The IHA had earlier estimated that global capacity includes at least 11,000 power stations and 27,000 generating units. Total electricity generation from hydropower was around 3983 TWh in 2014 according to the International Energy Agency. This is 16.4% of the global total electricity production in 2014 of 23,816 TWh.

[1]International Hydropower Association. Hydropower status report, 2017.

Hydropower. DOI: https://doi.org/10.1016/B978-0-12-812906-7.00001-6

Hydropower is widely distributed and only few regions are without significant hydropower potential. The countries of the developed world have exploited many of their best sites and hydropower generation forms part of the bedrock upon which developed the nations' prosperity is based. Development elsewhere has been slower but there have been major advances in Asia, particularly in China, in recent years and many of the countries of South America rely heavily on hydropower for electricity generation. Even so, most of these regions have much remaining capacity while in Africa hydropower is significantly underdeveloped.

In spite of its potential and obvious advantages, the development of hydropower can often be difficult, particularly where large projects are concerned. Major hydropower projects are often extremely disruptive and if not developed sensitively they can lead to a range of environmental problems. Large hydro-plants, particularly those involving dams and reservoirs, will inevitably change the environment in which they are constructed leading to displacement of people and wildlife and the destruction of ecologies. With care these changes can be managed but careless and sometimes reckless development during the second half of the 20th century led to hydropower acquiring a bad reputation during the 1980s and 1990s.

Since then the industry has made an effort to reform its practices and the World Commission on Dams addressed the main problems in *Dams and Development: A New Framework for Decision Making.*[2] This report proposed a complete reassessment of the criteria and methods used to determine whether a large hydropower project should be constructed. It also laid out an approach to decision-making which took account of all the environmental and human rights issues that a project might raise, an approach that will, potentially, filter out bad projects but allow well-conceived projects to proceed.

When projects are well designed and construction is carried out carefully, large hydropower schemes have the potential to transform the lives of those who benefit from them. Many such schemes provide water for irrigation and drinking as well as power and they can allow new industries to be established too.

---

[2]*Dams and Development: A New Framework for Decision Making*, the World Commission on Dams, Earthscan, 2000.

Economically hydropower is often considered expensive to build but when accounted for correctly it can become one of the cheapest sources of electricity available. Since 2000, the introduction of large quantities of renewable generation from wind and solar power has also led to the recognition that hydropower has an important role to play in the balancing of intermittent renewable generation on grid systems. This is leading to a further reassessment of the role of hydropower. Pumped storage hydropower plants, which are large-energy storage plants based on hydro-technology, can be used to store energy from renewable plants for use when needed. However conventional hydropower can provide significant grid support for other renewable generation too.

Large hydropower projects, above 30 MW in size, are not generally considered by regulatory authorities to be new renewable generation and in most regions do not attract support such as grants, special tariffs or tax breaks. However smaller hydropower schemes, which are generally classified as 'small hydropower', will often be included among the technologies that attract such support mechanisms. These smaller schemes are also less disruptive than their larger relatives and are consequently much easier to build.

THE HISTORY OF HYDROPOWER

The history of water power is inextricably linked to that of the water wheel although the latter was sometimes driven by manpower or by animals and used to raise water rather than using water to provide a source of mechanical energy. However some early water wheels such as the Noria wheel did exploit the flow of water to both drive a wheel and to raise water in pots attached to its circumference. This type of water wheel is believed to have been in use in the Roman Empire from as early as 700–600 BC.

The earliest use of water wheels may date back further still to 4000 BC. Babylonian inscriptions suggest that irrigation machines were in use in ancient Mesopotamia at around this period, although it is impossible to determine whether these involved water wheels. There is also a brief reference to water wheels in Sumeria (southern Mesopotamia) although again the precise interpretation is obscure.

Clearer evidence for water wheels exists in a ninth-century Arabic translation of Philo of Byzantium's treatise called *Pneumatica*, the original of which has been dated to between 240 and 220 BC.

In addition Mithridates VI, the ruler of Pontus in northern Anatolia, is recorded as having a water wheel in his palace in 71 BC. There is evidence from Egyptian papyri from the second and third centuries BC and from an Egyptian tomb fresco in Alexandria dated to the second century BC. Meanwhile the writings from the first century BC of the Roman engineer and architect Vitruvius contain the earliest clear description of a vertical water wheel. This is the Noria water wheel.

China saw early inventions during the Han dynasty which ended in AD 9; the first written Chinese evidence is contained in *Huan Zi Xin Lun* (New Discourses of Master Huan) which suggests knowledge of water wheels in the AD first century.[3] Here, as elsewhere, the initial uses of water wheels were to raise water for irrigation and for human consumption. However by the beginning of the first millennium they were also being exploited for milling as well as providing power for devices such as trip hammers that were used to break up ores.

Early civilisations are known to have used three different types of water wheel called the horizontal water wheel, the vertical undershot water wheel and the vertical overshot wheel. These are shown schematically in Figs. 1.1–1.3. The horizontal wheel, with a vertical shaft, can drive a millstone directly but in its simple form it is less efficient than the vertical wheel. The origins of each wheel design are unclear and evidence for all three exists in many parts of the world. There is some indication that Chinese developers preferred horizontal wheels whereas in the Middle East and Europe vertical wheels were more popular. However the earliest evidence for the horizontal wheel also comes from the Middle East and early vertical wheels can be found in China.

During the first two centuries of the first millennium, large mills based on water power were developed in both Europe and Asia. One of the largest mills, discovered in Barbegal near Arles in France in 1940, had 16 overshot water wheels. Large mills were also developed in China at around this time and water power was developed for crushing ores using a trip hammer. Meanwhile in Europe and the Middle East water wheels were often used to raise water; such wheels were driven by manpower or animal power. China appears to have relied more heavily on water power during the first millennium than

[3]Encyclopedia of Technology and Invention, Basic Books, New York, 2011.

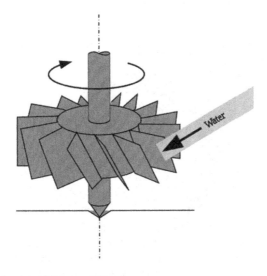

Figure 1.1 Horizontal water wheel. Source: Wikipedia.

Undershot water wheel

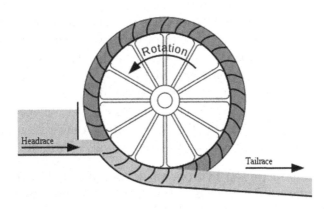

Figure 1.2 Vertical undershot water wheel. Source: Wikipedia.

Europe; after the fall of the Roman Empire the use of water wheels dwindled in the regions it formerly occupied.

The use of water wheels in Europe began to grow again during the medieval period. The reasons for this are not clear although several theories have been proposed including populations devastated by

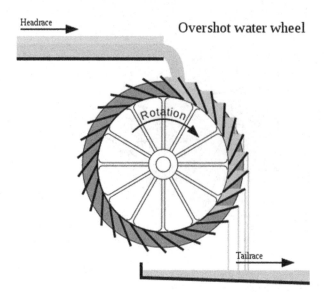

Figure 1.3 Vertical overshot water wheel. Source: Wikipedia.

plague unable to provide the manpower to drive wheels. Meanwhile history records that the number of mills was growing rapidly. A survey from the 10th century England recorded fewer than 100 mills. The Doomsday Book of 1086 lists 5624 water mills. The story was similar in France where one region recorded 14 mills in the 11th century and nearly 200 in the 13th century. This period also saw the development of the first tidal mills. The Cistercian monastic order helped drive the use of mills in Europe from the 12th century onwards, using water power for milling, wood cutting, forging and olive crushing.

The modern history of hydropower is usually dated to the publication of *Architecture Hydraulique* between 1737 and 1753 by the French engineer Bernard Forest de Bélidor. This book contained descriptions of both horizontal and vertical hydraulic machines. The industrial revolution, which began in England in the 18th century, was originally driven by hydraulic power. This was gradually superseded by steam power but many water-powered devices continued to be in use in the 19th century. It was during the 19th century that water power was first harnessed to generate electricity; hydropower was born.

Advances in the understanding of hydraulic technology led to the enclosing of water wheels in a casing to improve efficiency; these new devices have come to be known as water or hydropower turbines. The first device to bear this name was the Fourneyron turbine designed by the French engineer Benoît Fourneyron. His first unit was a 6-horse-power (4.5 kW) device built in 1827 but by 1837 he had designed a 45-kW turbine capable of 2300 rpm and operating with 80% efficiency. The design, which was used for the hydropower plant at Niagara Falls in 1895, is still in use today.

Other designs followed. In 1849 the American engineer James Francis developed the Francis reaction turbine, one of the most widely used turbine types for hydropower plants. In 1889 another American engineer, Lester Allen Pelton, patented his Pelton turbine design, now used widely for high head hydropower applications. The Kaplan turbine, a propeller turbine with adjustable blades, was developed in 1913 by Austrian scientist Viktor Kaplan. All that was needed now was to couple one of these new turbine designs to an electric generator, which was also a revolutionary new device, to start the age of electricity.

That step was accomplished by the English engineer and entrepreneur William Armstrong who, in 1978, built and operated the first electric power station at his home at Cragside, Northumbria. Armstrong used hydro-turbines driven by a river on his estate to drive a Siemens dynamo, a device that generates direct electric current. The power station was first used to provide lighting at Cragside but was later employed for a range of other services.

Other early developments include two US power plants using dynamos driven by water turbines, one built in 1880 in Grand Rapids, Michigan, which supplied arc lighting to a theatre and a store front and a similar plant built in 1881 in Niagara Falls, New York, which supplied street lighting. In 1882 a paper manufacturer, H. J. Rogers, built a plant at Appleton Wisconsin which provided electricity to light Roger's home and his paper plant. This plant had an output of 12.5 kW. By 1889 there were estimated to be 200 hydropower plants in the United States. These early plants all generated direct current. The first alternating current hydropower plant in the United States was the Redlands power plant in California, which began operating in 1893.

The use of hydropower spread quickly. A publicly owned plant was built in Australia in 1895. In 1905 a station was built near Taipei with a generating capacity of 500 kW and construction soon followed on mainland China. Meanwhile, rapid development continued in North America with the Hoover Dam, built in 1936, having a generating capacity of 1345 MW. Later in the century the momentum moved to Asia and South America. The two largest hydropower plants in operation today are the 22,500-MW Three Gorges dam in China, which was opened in 2008 and the 14,000-MW Itaipu power station on the border between Brazil and Paraguay, which started operating in 1984.

GLOBAL PRODUCTION OF ELECTRICITY FROM HYDROPOWER

Hydropower is the most important global renewable source of electricity. Since its beginning, at the end of the 19th century, it has supplied a significant part of the world's electrical power. Detailed production figures for the first half of the 20th century are not available but figures from the IEA indicate that between 1970 and 2015 hydropower has contributed between 20.3% and 15.4% of the annual global electricity production. Production from hydropower plants is irregular because it depends on global rainfall and some years are more productive than the other. However the global contribution as a percentage of total output has fallen slowly over the past 60 years even though absolute production has risen.

Absolute production figures indicate that total hydropower production was over 920 TWh in 1965 based on figures from BP.[4] This had risen to 1296 TWh by 1973, based this time on figures from the IEA and shown in Table 1.1. Production has generally risen year on year since then although as noted above there are years when rainfall is low and this reduces output. A continuous set of hydropower production figures from 2004 to 2014 are also shown in Table 1.1. As these indicate, production in 2004 was 2808 TWh, more than twice that in 1973. By 2006 global production was 3121 TWh and by 2010 it had reached 3516 TWh. in 2015 total global production was 3978 TWh. Separate figures from BP indicate that total production exceeded 4000 TWh in 2016.

[4]BP Statistical Review 2017. The figure in the text is based on consumption figures from the BP report.

Table 1.1 Annual Global Hydropower Production	
Year	Global Hydropower Production (TWh)
1973	1296
2004	2808
2005	2994
2006	3121
2007	3162
2008	3288
2009	3229
2010	3516
2011	3566
2012	3756
2013	3874
2014	3983
2015	3978
Source: *IEA[5]*.	

Although most countries in the world have some hydropower generating capacity, global production is dominated by a small number of very large producers. Since the beginning of the 21st century the largest producer has been China which has expanded its capacity rapidly in recent years. In 2015 the country produced 1130 TWh of electrical power from its hydropower plants, 28.4% of the global total hydropower output, as shown in Table 1.2. The world's next largest producer of hydropower was Canada with an output of 381 TWh in 2014, 9.6% of the global total. Brazil, with 360 TWh and the United States with 271 TWh make up the top four nations which between them accounted for 53.8% of all global hydropower production in 2014.

The remaining nations each account for less than 5% of global production. The most important of those are Russia with 170 TWh (4.3%), Norway with 139 TWh (3.5%), India with 138 TWh (3.5%), Japan with 91 TWh, Sweden with 75 TWh and Venezuela with 75 TWh. A further 28.8% of global production is spread across the rest of the globe.

Table 1.3 focuses on the global installed hydropower capacity. According to the IHA this was 1,345,500 MW at the end of 2016. Table 1.3 shows this total broken down by region. East Asia and the

[5]IEA Key World Energy Statistics 2006–2016.

Table 1.2 Global Hydropower Production by Country, 2015		
Country	Hydropower Production (TWh)	Percentage of Total
China	1130	28.4
Canada	381	9.6
Brazil	360	9.0
United States	271	6.8
Russia	170	4.3
Norway	139	3.5
India	138	3.5
Japan	91	2.3
Sweden	75	1.9
Venezuela	75	1.9
Rest of the world	1148	28.8
Total	3978	100.0
Source: IEA[6].		

Table 1.3 Global Installed Hydropower Capacity by Region, 2016	
Region	Installed Capacity (MW)
Africa	33,624
East Asia and the Pacific	457,473
Europe	223,008
North and Central America	200,922
South America	164,071
South and Central Asia	166,402
Total	1,245,500
Source: IHA[7].	

Pacific, dominated by China, has the largest regional total with 457,473 MW of hydro-generating capacity at the end of 2016. Europe has the second largest total at 223,008 MW, closely followed by North and Central America with 200,922 MW. South and Central Asia accounted for 166,402 MW while South America was home to 164,071 MW. Africa had the smallest capacity at the end of 2016, with 33,624 MW.

Table 1.4 shows figures for the top 20 nations, rated by their installed hydropower generating capacity in 2016. The order more or less mirrors that of Table 1.2 but some nations have swapped positions, indicating

[6]IEA Key World Energy Statistics 2017.
[7]International Hydropower Association Hydropower status report 2017.

Table 1.4 Top 20 Countries by Installed Capacity, 2016	
Country	Installed Capacity
China	331,110
United States	102,485
Brazil	98,015
Canada	79,323
India	51,975
Japan	49,905
Russia	48,086
Norway	31,626
France	26,249
Turkey	25,405
Italy	21,884
Spain	20,354
Switzerland	16,657
Sweden	16,419
Vietnam	16,306
Venezuela	15,393
Austria	13,177
Mexico	12,092
Colombia	11,606
Germany	11,258
Source: IHA[8].	

that average output per megawatt varies from country to country (and potentially from year to year). China heads the table with 331,110 MW installed at the end of 2016, followed by the United States with 102,485 MW and Brazil with 98,015 MW. Canada had 79,323 MW, India 51,975 MW and Japan 49,905 MW, followed by Russia with 48,086 MW and Norway with 31,626 MW. Other nations in the table include European countries France, Italy, Turkey, Spain, Switzerland, Sweden, Austria and Germany. From the Americas there are Mexico, Venezuela and Colombia and from Asia, Vietnam.

Finally, Table 1.5 shows the global capacity for pumped storage hydropower, broken down by regions. Pumped storage hydropower is an important energy storage technology that allows large quantities of

[8]International Hydropower Association Hydropower status report 2017.

| Table 1.5 Global Installed Pumped Storage Hydropower Capacity by Region, 2016 ||
Region	Installed Capacity (MW)
Africa	3376
East Asia and the Pacific	64,684
Europe	50,467
North and Central America	22,618
South America	1004
South and Central Asia	7541
Total	149,690
Source: IHA[9].	

electrical energy to be stored and then returned to the grid when it is needed. Many early plants of this type were built to support nuclear power because nuclear plants often operate most efficiently if they can run at full power all the time. So, when demand falls at night, surplus energy from these plants is stored for use the next day. More recently energy storage has become important as a way of balancing the production and use of energy from intermittent renewable resources on the grid.

At the end of 2016 the total global capacity of pumped storage hydropower was 149,690 MW. The largest regional capacity was in East Asia and the Pacific with 64,685 MW, followed by Europe with 50,467 MW and North and Central America with 22,618 MW. The other three regions in the table each has less than 10,000 MW.

[9]International Hydropower Association Hydropower status report 2017.

CHAPTER 2

The Hydropower Resource, Hydropower Sites and Types of Hydropower Plants

The energy that is extracted from water by a hydropower plant and converted into electricity is potential energy contained within the mass of water as a consequence of its elevation. This elevation is normally calculated with reference to sea level which is the lowest level to which the water can normally flow. The energy contained within the mass of water flowing down a river is released as the water flows downhill, normally being dissipated in various ways within the watercourse down which it flows as well as being carried by the water itself as kinetic energy. Turbulence, erosion, noise and the picking up and carrying of particles of silt are all evidence of the energy being dissipated. A hydro-turbine can extract some of this energy and use it to produce electric power.

The initial elevation of water that flows down a water course is a result of solar energy heating surface water in bodies such as seas and lakes, causing evaporation of water into the atmosphere as water vapour. Once solar energy has heated the surface water and caused evaporation, the water vapour gets carried high into the atmosphere where it eventually forms clouds. Movements and pressure changes within the atmosphere then carry the clouds around the globe and eventually much of the water within them is returned to the earth's surface as rain. It is this rain that feeds all the world's water courses and provides the energy that can be exploited by hydropower plants.[1] Ultimately, therefore, hydropower can be considered as a form of solar energy.

How much energy is available as a result of the water flowing down a region's or the world's watercourses? The answer is a large amount, but making an accurate estimate is extremely difficult. There are two possible approaches, top-down and bottom-up. The bottom-up

---

[1]There is also some melt-water from the world's glaciers which are gradually melting as the world warms. The original source of the ice in glaciers is rain.

Hydropower. DOI: https://doi.org/10.1016/B978-0-12-812906-7.00002-8

approach is to examine all the waterways in a region and estimate the amount of energy that might be extracted from each, then add all the amounts together. The top-down approach, meanwhile, is to examine how much rain falls across a region and try and estimate the potential energy that contains once it reaches the earth.

When looking at the global resource, it is common to use the top-down approach. A number of different measures can be derived using this technique. The first is the gross theoretical hydropower potential of a country, region or the world. The gross theoretical hydropower potential of a region is defined by the World Energy Council (WEC) as 'the annual energy potentially available in the country if all natural flows were turbined down to sea level or to the water level of the border of the country (if the watercourse extends into another country) with 100% efficiency from the machinery and driving waterworks'.[2] The energy available is a function of both the mass of water and the distance it falls after it has reached the earth's surface. Carrying out the calculation to sea level maximises the energy available. It is not an easy calculation to make and involves estimates and compromises but it does provide figure upon which to base other estimates. However the WEC itself warns that the figures should be used with caution.

Table 2.1 contains figures for the gross theoretical hydropower capacity for all the major regions of the world, published by the WEC. Based on these figures, Asia has the greatest potential, put at

Table 2.1 Regional Hydropower Potential		
Region	Gross Theoretical Hydropower Capacity (TWh/year)	Technically Exploitable Hydropower Capacity (TWh/year)
Asia	16,618	5590
Europe	4919	2762
North America	5511	2416
South America	7541	2843
Africa	3909	1834
Oceania	654	233
Middle East	690	277
World total	39,842	15,955
Source: World Energy Council[3].		

[2] 2004 Survey of Energy Resources, World Energy Council.
[3] 2010 Survey of Energy Resources, World Energy Council.

16,618 TWh/year, followed by South America with 7541 TWh/year, North America with 5511 TWh/year, Europe (4419 TWh/year) and Africa (3909 TWh/year). Potential in the Middle East (690 TWh/year) and Oceania (654 TWh/year) is much more limited.

Based on the figures in Table 2.1, the global gross theoretical hydropower potential is 39,842 TWh/year. Other sources have also made estimates. For example, a study by Eurelectric published in 2000 suggested that the global total was around 51,000 TWh/year.[4] As a comparison, total global electricity production in 2015 was 24,255 TWh according to the IEA.[5]

There are technical, economic and environmental reasons why this gross capacity can never be fully realised. Some of the flowing water is inaccessible, some is in regions where building a hydropower plant would be too disruptive and some is simply too difficult to harness. A second measure, the technically exploitable hydropower capacity, provides a more realistic figure for the amount that might eventually be used. This is a measure of the capacity that could be exploited using currently available technology and its estimation is often based on site visits. Regional technically exploitable hydropower capacities are also shown in Table 2.1. These are significantly smaller than the gross theoretical capacities. Across Asia, the technically exploitable capacity is 5590 TWh/year, 34% of the gross theoretical capacity. Technical capacities in other regions are similarly reduced in comparison to the gross capacity. Total global technically exploitable potential is put at 15,955 TWh or roughly 40% of the theoretical total.

Cost may further reduce the potential capacity since some technically exploitable hydropower sites will be too costly to develop. To take account of this, a third measure of potential capacity is the economically exploitable hydropower capacity. The global economically exploitable hydropower capacity has been estimated to be about 13,100 TWh/year.[6] There may be still further limits on development as a result of environmental concerns or other restrictions. One final measure of hydropower potential, the exploitable hydropower capacity,

---

[4]Jóhann Már Maríusson, J. M., & Thorsteinsson, L. (2000). *Study on the importance of harnessing the hydropower resources of the world.* Eurelectric.
[5]IEA Key World Energy Statistics 2017.
[6]This estimate is from Eurelectric.

reflects this. The global exploitable hydropower potential is around 10,480 TWh/year.

As noted in Chapter 1, global hydropower generation in 2015 was 3978 TWh or around 38% of the exploitable potential. On this basis, just over 60% of exploitable global potential remains to be developed. However while such estimates are useful guides, all these potential figures should be treated as approximations because there is no general consensus about how such estimates should be made.

Another useful guide to regional hydropower potential is to look at the regional level of hydropower exploitation or the proportion of the estimated technical hydropower potential in each region that has already been exploited. Based on the technically exploitable potential figures from the WEC shown in Table 2.1, the most highly developed region is Europe where 26% of the technically exploitable potential has been developed. Europe was an early exploiter of its hydropower potential and by some estimates most of the best hydropower sites have now been used. Even so these figures suggest that there remains significant potential to harness. The next highest level of exploitation is in North and Central America with 22%. South America has exploited 20% of its technical potential and Oceania 21% but Asia has only exploited 13%, the Middle East 10% and Africa 5%. These figures suggest that there is a large amount of hydro-capacity that could still be built in most regions of the world. However the region that stands out the most is Africa which has barely touched its technical potential. Given the economic and developmental problems across many countries in Africa, hydropower would appear to offer a great opportunity.

An alternative analysis of global hydropower potential has been carried out at the Delft University of Technology. The results suggest that there may be even more potential to exploit than the figures from the WEC cited above suggest. This study used satellite mapping data and computer modelling to simulate the total global hydropower capacity. This can potentially reveal hydropower capacity that has not been included in the technical capacity figure above because regions are difficult to access. Figures from this analysis are shown in Table 2.2.

Table 2.2 Global and Regional Hydropower Potential Generating Capacity[a]				
Region	Total Potential (GW)	Large Hydropower Potential (GW)	Small Hydropower Potential (GW)	Micro Hydropower Potential (GW)
Africa	2111	1729	355	27
Asia	13701	11,714	1775	212
Oceania	52	7	30	15
Europe	792	540	217	35
North America	2038	1653	327	58
South America	1296	1001	245	50
World	19,990	16,644	2949	397

[a]Totals have been corrected where there were inconsistencies.
Source: Delft University of Technology[7].

The figures in Table 2.2 are for generating capacity rather than energy production. As such, this provides a different measure of global potential. Based on this analysis, the region with the largest potential is Asia where a total potential is estimated to be 13,700 GW. Potential capacity in other regions is much smaller. In Africa it is 2112 GW, in North America the figure is 2038 GW, South America could provide 1251 GW and Europe, 792 GW. Not surprisingly given the size of the landmass, Oceania has the lowest potential, 52 GW. One particularly valuable insight provided by this analysis is the extensive unrealised hydropower potential of several countries including Colombia, Myanmar, Indonesia and Madagascar. Again for comparison purposes, the total global installed hydropower generating capacity at the end of 2016 was 1246 GW according to the IHA.[8]

HYDROPOWER SITES

It may appear to be stating the obvious but in order to build a hydropower plant the first thing that is required is a suitable site. It is important to realise at the outset that hydropower is extremely site specific. A major river may be hundreds of kilometres long but there may only be a few places where a successful power plant can be built. What is more, the type of hydropower plant that can be built will depend on

[7]World Hydropower Capacity Evaluation, L. J. J. Meijer, R. J. van der Ent, O. A. C. Hoes, H. Mondeel, K. E. R. Pramana and N. C. van de Giesen. Available from Delft University of Technology archive https://repository.tudelft.nl/islandora/object/uuid:49d8d013-9a48-4222-85a1-0ebb53f83dc8/?collection=research.
[8]International Hydropower Association Hydropower status report 2017.

the topography of the site, once chosen. Factors such as the terrain and subterranean rock structure, the incline of the riverbed at the site and the size of the river will all affect if, and how it can be exploited. If the proposed project is large, then access will be important. Some of the best hydro-sites are in extremely challenging positions that make construction difficult and therefore costly. The environmental impact will have to be taken into account too.

A successful hydropower project requires a river with suitable hydrological conditions too. The amount of energy that can be taken from any river will depend on two factors, the volume of water flowing along it and the drop in riverbed level, normally known as the head of water that can be exploited by a plant at a particular site. The available power increases with the volume of water − the bigger the river, the more power available − while a steep riverbed carrying a fast flowing river will generally yield more electricity than a slowly descending, sluggish one of similar size.

For a given volume of water, the energy available will depend directly on the head height or drop in water level that can be utilised and this is normally larger the steeper the riverbed. This does not mean that slow-flowing rivers are not suitable for hydropower development. They often provide sites that are cheap and easy to exploit. In contrast, steeply flowing rivers are often in inaccessible mountainous regions where construction is challenging.

The capacity available at a hydropower site can vary from a few kilowatts to many hundreds of megawatts. Occasionally, sites will yield thousands of megawatts. The largest single developed site in the world is the Three Gorges dam on the Yangtze River, China, with a generating capacity of 22,500 MW. Probably the largest unexploited site is on the Congo River in Africa where a multiple barrage development is estimated to be capable of supporting up to 35,000 MW of generating capacity. This is exceptionally large; most are smaller. Large projects of this type, where developed, are likely to be multipurpose projects involving flood control, irrigation, fisheries and recreational usage as well as electricity generation. They are often disruptive and the capital outlay for construction is large. With multiple uses, funding becomes more complex too. Smaller projects may be multipurpose of they may simply generate electricity. Some sites are tiny, suitable to support a few households but they can be very cheap and easy to exploit.

In choosing a site, hydrology is important but so too are geography and geology. Given a river capable of supplying energy, the optimum site or sites for extracting this energy will be determined by the geography. A steep-sided valley may offer the opportunity to build a dam, an opportunity that more open ground cannot provide. On the other hand the more open site may be easier to exploit and closer to the demand centres where the power is needed.

Once a site has been identified, an extensive geological survey will then be necessary to determine the underlying structures. Many hydropower plants are physically massive and can generate enormous pressures, leading to stresses in underlying strata and potential fractures. These can be disruptive if possible faults are not identified before work begins. Where large reservoirs are involved, more stress can be created by the mass of water impounded by a dam and this has in some cases led to the generation of seismic tremors as underlying strata react. The effects are usually small but they are, nevertheless, alarming and potentially damaging.

There are often environmental issues too. A large hydropower development with a dam might involve flooding a massive area within a river valley. Habitat will be lost, people may be displaced and cultural or religious sites may be inundated. In some cases rare or endangered species are involved. All these must be evaluated before a project can proceed. In many cases work will have to be carried out to mitigate the effects of the project, re-housing people, creating a new environment for displaced animals, moving cultural sites or artefacts. In some cases it may be judged too disruptive and authorisation to construct will be withheld.

Given that hydropower depends on having a suitable site, how does one set about locating a hydropower site? Much will depend on the region. Many countries have carried out at least cursory surveys of the hydropower potential within their territory and provisional details of suitable sites are available from the water or power ministries. Global satellite surveys can also help identify suitable sites. Sometimes much more detailed information is available too. However all these types of data are superficial in nature and cannot replace an on-site survey. Indeed surveys carried out as part of a feasibility study form an integral of any hydropower scheme. For a large scheme a feasibility study may account for 1% or 2% of the total cost. For smaller schemes it can reach up to 50%.

Site surveys should examine the geography, the geology and the hydrology of the site. The latter is particularly important as it will reveal how much power the plant can expect to produce and how it will vary with the seasons. In some cases there will be hydrological data available that can reveal how much water flows down the river and when, revealing seasonal variations. If this type of data is not available then it must be collected. However it will be impossible to amass years of data in this way so historical research will be necessary to try and establish how flow has varied, year on year. Also vital is to determine the likely maximum flow during a period of flooding. Underestimating the likelihood of a major flood or its size could lead to the ultimate failure of the project, either before or after it has been built. Most projects will be built to withstand flood at least one in 100 years. Underestimating the potential size of such an event can lead to disaster.

CATEGORIES OF HYDROPOWER PLANT

As has already been highlighted, hydropower plants come in a wide range of sizes. The largest have generating capacities of thousands of megawatts, the smallest a few kilowatts. The techniques used to build these plants vary too. Large plants will always be built to suit a particular site and all the components are likely to be tailored for this specific project. Smaller plants, in contrast, will often use off-the-shelf components to keep costs down.

Given these variations, hydropower plants are traditionally broken down into categories depending on their size. Categories can vary, depending on the defining agency but the most common categorisation is shown in Table 2.3. The smallest plants, with capacities of between

Table 2.3 Hydropower Plant Categories	
Micro	1 kW–100 kW
Mini	100 kW–1 MW
Small	1 MW to 10–30 MW
Large	Above 10–30 MW
Source: *UNDP/World Bank*[9]	

[9]Private mini hydropower development study: the case of Ecuador, UNDP/World Bank, 1992.

1 kW and 100 kW are called micro hydropower plants. Between 100 kW and 1 MW a plant is described as a mini hydropower plant. Small hydropower plants are generally those with capacities of between 1 MW and 10 MW but this upper limit can vary from country to country and in some cases may be as high as 30 MW. Plants with capacities larger than 10 MW (or up to 30 MW depending on jurisdiction) are classified as large hydropower plants.

Sometimes an intermediate category for medium hydropower plants is also introduced between small and large hydropower. If used, this is typically for plants between 5 and 50 MW; those above are large and those below are small. From a global perspective, large hydropower is the most important category and this accounts for most of the hydropower capacity in operation today. However the smaller categories are important locally.

Large hydropower plants are the most technically sophisticated and are generally individually designed for each site using turbines that have also been made specifically for the power plant. Small hydropower plants are similar to large plants but some use off-the-shelf turbines and other components rather that bespoke components. Mini and micro hydropower installations usually employ standard turbines and many involve novel, often cost-effective, designs not used in larger plants.

Table 2.2 provides an indication of the potential for these different categories of hydropower plant across the globe. In this table large hydropower is defined as being greater than 10 MW, small hydropower covers 100 kW to 10 MW and micro hydropower is below 100 kW. As would be expected, large hydropower potential dominates in most regions. Globally it accounts for 16,644 GW out of a total of 19,343 GW or 83% of the total. Small hydropower is estimated to potentially offer a further 2949 GW and micro hydropower a further 396 GW. It is worth noting, however, that small hydropower offers the largest potential in Oceania, 30 GW out of a total of 52 GW. In Europe too, the small hydropower contribution is large, with 217 GW (27%) out of a total of 792 GW.

CHAPTER 3

Dams and Barrages

In order to extract energy from flowing water in a hydropower plant, the water from a waterway must pass through a turbine, forcing the machine to rotate and drive a generator. For small and very simple hydropower plants, it may be sufficient to place a suitable turbine into the moving water. Old style water mills with paddle wheels often employed this approach and more modern small propeller turbines can be deployed in this simple way too. In most cases, however, this does not make the best use of the energy available.

If a device is simply inserted into the flow only a part of the available energy will be harvested. In order to extract as much energy as possible it is necessary to use all the water that is available. This involves physical intervention into the waterway to divert and manage the flowing water. Many old water mills managed the flow of water with mill races that controlled the flow of water to the mill wheel. Modern hydropower plants will usually do the same.

In the example of a water mill, the water in a stream or river is diverted so that it passes through a narrow channel, the mill race, where the mill wheel is positioned.[1] Allowance must be made for when the river is in flood; this requires a provision so that some or all the water can bypass the mill wheel channel, as necessary. There may also be a mill pond to collect and store an amount of water so that some is always available to drive the mill wheel. Like the suspension of a car smoothing a roadway, this also irons out some of the peaks and troughs in the water flow.

Modern hydropower plants designed to generate electrical power use similar strategies and the water mill provides an example of two of the most important ways of harnessing the water in a river or stream. The first and perhaps the most important is the dam and reservoir power plant. In this type of hydropower plant a large physical

---

[1]A mill race may only take part of the water from the waterway.

Hydropower. DOI: https://doi.org/10.1016/B978-0-12-812906-7.00003-X

structure, a dam, is built across the river. Once in place, this dam will cause a lake — the reservoir of water — to build up behind it. As its name implies, the reservoir provides a reserve of water that is stored and ready to be used to generate power as necessary while also damping the peaks and troughs of the flow. Channels within the dam allow water to be taken from the reservoir and delivered to one of more hydro-turbines which each provide the mechanical power to drive a generator and provides electrical energy. Once through the turbine(s) the water is released back into the original waterway.

Building a dam and creating a reservoir provides an extremely effective way of managing the water flow in a river and harnessing it for electricity production but it is also extremely disruptive both during construction and afterwards, in operation. The dam itself will be a massive structure in the landscape while the reservoir that builds up behind it will submerge a large area of land. This reservoir will certainly even out the water flow, capturing water during the flood or rainy season and releasing it during dryer periods but this will affect the amount of water downstream and may destroy habitats that have relied on the previous seasonal flow.

The alternative is to do away with the dam and reservoir in favour of a simpler solution. This is the basis for the second important type of hydropower scheme called a run-of-river project. They key to this kind of hydropower development is that water is not dammed but just diverted from the flowing river and passed through a hydropower plant, then returned to the river afterwards. The diversion usually requires a low barrage to be built across the river but this is simply to control and divert the flow, not impede and collect it. Such a scheme is much less disruptive of the riverine environment than a dam and reservoir. It is also less effective because it relies entirely on the flow of water in the river for its energy supply. There will be more water during the rainy season and less during the dry season and output from the plant will vary accordingly. Moreover during high flow periods, some of the water may have to be spilled past the power plant, losing the energy it contains.

All large and most small hydropower plants will conform to one of these two types. More modestly sized plants may adopt other designs such as the insertion of a turbine into the flow typical of the simple water mill described above. In addition, mini and micro hydropower plants often utilise novel designs such as inflatable rubber barrages that could not be used on larger developments. However all except the simplest will usually attempt to manage the water flow in some way.

RUN-OF-RIVER HYDROPOWER

A run-of-river hydropower scheme is a type of power plant that takes the water flowing down a waterway, diverts it so that it passes through a hydro-turbine generator to generate electric power and then returns the water to the waterway at a point lower down the watercourse. The water is usually diverted by placing a low barrage across the river or stream and then extracting water from the pond that collects behind it. In most cases this type of power plant will have no significant water storage but some plants do incorporate a small storage volume which is often referred to as the pondage[2] of the plant. This will generally provide little flow regulation so the plant can still be classified as run-of-river. However, a limited pondage may allow a scheme to vary its output on a daily basis depending on the demand.

As a consequence of its relative simplicity, a run-of-river hydropower scheme is usually the cheapest hydropower project to develop. Since it requires no dam, a major constructional cost and civil engineering project is avoided. The geological problems that are often associated with dam construction are avoided too.

Any hydropower plant requires a head of water that can be exploited if energy is to be extracted. In a dam and reservoir plant, the dam can itself provide the head of water. With a run-of-river plant this is not possible and so another means of creating a suitable head must be found. This is normally achieved by choosing a point significantly lower down the waterway from the diversion structure to build the power house. The difference in height between the diversion structure and the power house can then be exploited to provide the necessary head.

In order for this to operate effectively, the water from the diversion structure must be carried to the power station and its turbines with minimum energy loss. This is best achieved by first diverting the water from the river into a canal called the headrace that carries the water to a spot close above the power house of the plant. The headrace is

----

[2]In this context the pondage is the volume of water in the 'pond' behind the barrage.

normally constructed in a way that exploits the local topology in order to achieve the minimum drop in elevation along its length because any drop in height is a loss of energy.

Once the water reaches the end of the headrace, it falls steeply through a pipe called the penstock into the power house where it is directed into the turbine(s). The design of the penstock will seek to avoid unnecessary turbulence that can lead to further loss of energy. (In smaller projects the headrace and penstock may be the same pipe, as shown in Fig. 3.1). The height of the penstock intake above the power house represents the head of water that is available for generating power in the plant. The head of water generates a pressure of water at the bottom of the penstock and it is this pressure, dependent upon the height of that head, which provides part of the force to drive the turbines. The kinetic energy of the fast flowing water will also help

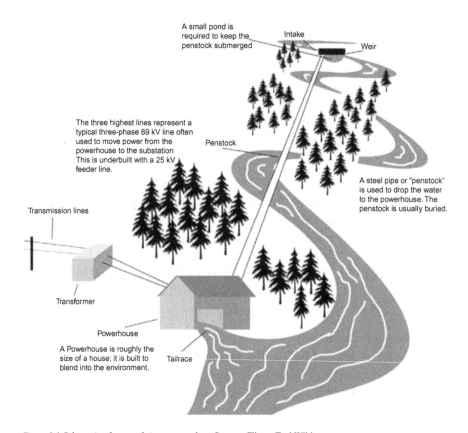

Figure 3.1 Schematic of a run-of-river power plant. Source: ClimateTechWiki.

turbine the machines. Having passed through the turbines the water is returned to the river at a point downstream of the diversion structure through another pipe called the tailrace.

The length of the various channels used to divert the water for a run-of-river plant are kept as short as possible but much depends on the topography of the area where the generating station is built. For large run-of-river plants the headrace/penstock can sometimes be tens of kilometres long in order to achieve a suitable head of water. This can lead to a large stretch of the waterway being deprived of water when the plant is operating. To avoid this it may be necessary to let some water spill past the diversion structure in order to support downstream habitats.

The simplicity of the run-of-river scheme is attractive but it is also the main weakness of this type of development. With no dam to conserve water, the power plant must rely exclusively on the flow of water in the river. As this fluctuates, so will the amount of power that can be generated. Under drought conditions the plant will be able to generate no power, whereas when the river is in flood, much of the available water will have to be allowed to flow past the diversion system without being exploited. The same applies when power from the plant is not needed for the grid. Nevertheless this type of project does have significant advantages besides cost, particularly because of the relatively small amount of environmental disruption it causes.

Run-of-river hydropower plants are typically between 10 and 1000 MW in generating capacity. They could be larger, in theory but in practice larger plants of this type have not been built. However, a series of run-of-river power plants along the same river can exceed 1000 MW in capacity. There are also much smaller run-of-river schemes. These often use simple diversion structures and off-the-shelf materials to construct their headraces and penstocks in order to remain cost-effective.

DAM AND RESERVOIR PROJECTS

The alternative to the run-of-river configuration is a dam and reservoir project. This will involve a major civil engineering undertaking, the construction of a dam. There are a range of dam designs, each suited to a particular type of terrain. However all require careful engineering and project management.

The purpose of a dam is to create a reservoir of water which builds up behind it. The reservoir is essentially a form of energy storage system. Once created, the reservoir allows some measure of control over the flow of water in the river beyond the dam and consequently the flow through the turbines in the power house. Water can be conserved during periods of high flow and used up when rainfall is low. A dam can also be used for flood control, preventing damaging surges of water lower down the watercourse. More recently, dam and reservoir hydropower systems have been used to help balance intermittent forms of renewable energy on the electricity grid. This again relies on their ability to store energy in the form of water for use when necessary. The fact that hydro-turbines can respond quickly to changes in demand facilitates this application.

If a dam is to be constructed, then a very careful geological survey of the underlying soil and rock structure will be needed to identify any faults that might make a dam unstable or allow water to flow beneath or around it. Geological faults or unsuitable substrata need not prevent construction of a dam because they can be treated but if they are only discovered during construction, or later, they are likely to result in massive additional costs and delays. If they are not discovered, then there may be eventual catastrophic failure.

The layout of a dam and reservoir scheme is essentially the same as for a run-of-river plant. Water is extracted from the dam and carried though a headrace until it is above the power house where it enters the penstock and falls into the turbines. From the turbine hall it is then returned to the river through a tailrace. However, since the dam itself may generate the main head of water, the design can often be more compact than for a run-of-river plant.

The construction of a dam is a complex engineering project. Water flowing down the river must be temporarily diverted or coffer dams must be erected to isolate parts of the riverbed so that construction can take place. The size and complexity of dam construction means that this part of the project will account for up to two-thirds of total project costs.

While the dam forms the major part of the construction project, the reservoir behind the dam is likely to have the largest environmental impact. The lake created behind the Three Gorges dam is 600 km long but even small dams can create large areas of water with dramatic effects on the local environment. Once the water has been impounded behind the dam, it must be controlled so that it does not overtop the dam during flood conditions. Meanwhile the reservoir must be periodically flushed to remove sediment that is swept down the river; otherwise the dam may eventually fill with silt and become useless.

DAM TYPES

The dam in a hydropower project creates a barrier across a waterway and holds back the flowing water, creating a lake or reservoir behind the structure. Depending on its size, the dam will provide a range of other facilities too. One of the most important is a spillway through which water can be released from the reservoir if the latter becomes too full. Water should never be allowed to flow over the top of a dam as this could eventually lead to its failure. For a large dam on a navigable waterway there may be ship locks to allow shipping to pass the dam. Passages that allow fish to pass are often needed too, particularly on rivers in which migratory fish spawn. Depending upon the design of the hydropower scheme, there may be a powerhouse with turbine inside the dam structure. There will be inspection galleries too. And some dams also provide roadways across the river they are damming.

There are several different types of dam, each suited to a different situation. Four principle types are used for large and small hydropower plants: embankment dams, gravity dams, buttress dams and arch dams. In addition, there are three other types that have been used for smaller projects: steel dams, timber dams and rubber dams.

Embankment dams: An embankment dam is made from natural materials and they are usually constructed using material sourced locally. Ideally the material will be taken from land that will be submerged by the reservoir once it has formed. The dams come in two principle varieties, earthfill embankment dams and rockfill embankment dams. An earthfill embankment dam is built around a core of an impermeable material such as clay or concrete which is keyed into the rock of subsurface to prevent water flowing beneath it. This core is then buttressed by banks of earth, on both sides, to create a gradually

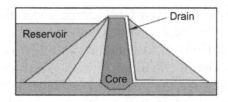

Figure 3.2 Cross section of an embankment dam. Source: The British Dam Society.

sloping wall on both the upstream and the downstream face of the dam (see Fig. 3.2). The earth provides the mass that holds the dam in place and resists the pressure from the water in the reservoir. Apart from the core the dam is permeable to water and care must be taken to control any water seepage through the earth structure. A rockfill dam is similar to an earthfill dam but the body of the dam is constructed from rock. This is also permeable to water and a rockfill dam may have an impermeable core, like an earthfill dam. Alternatively it may have its upstream face sealed with concrete to render it impermeable. One of the main dangers with an embankment dam is that overtopping of the dam can lead to erosion of the material from which it is constructed and to eventual failure. To prevent this, the structure will be fitted with a concrete spillway to control the water level. This type of dam can be built on both rock and soil substructures since it does not impose much pressure on its foundations because it is relatively light and shallow compared to other types of dam. The dams are usually used when a wide, shallow valley is to be dammed for power generation. They create wide, shallow reservoirs.

Gravity dams: A gravity dam is a massive dam constructed from concrete, masonry or both. As its name implies, it holds in position as a result of gravity acting on its mass; i.e it is held in position by its weight. Gravity dams normally have a vertical or near-vertical upstream face, while the downstream side is sloping to create a triangular structure, as shown in Fig. 3.3. This type of dam must be built on secure rock foundations because of the pressure it exerts on those foundations. It will usually be keyed into the rock both to increase its resistance to the pressure from the water and to prevent any water seeping underneath the dam. The dams can be built in both wide and narrow valleys and they can be much taller than embankment dams. The tallest concrete gravity dam, the Grande Dixence dam in Switzerland, is 285 m high.

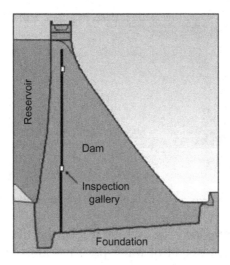

Figure 3.3 Cross section of a gravity dam. Source: British Dam Society.

Buttress dams: The buttress dam, sometimes called a hollow dam, is a variant of the concrete gravity dam. It too is built from concrete and has a vertical upstream face but downstream it is supported by massive triangular buttresses walls built into the rock below. The buttresses help the dam resist the pressure of the water on the structure. There are spaces between the buttresses, spaces that would be filled in the case of a conventional gravity dam hence the name hollow dam. In consequence, these dams require much less material than a conventional gravity dam and this has made them attractive. Like a gravity dam, a buttress dam can be built across both narrow and wide river valleys and like the gravity dam it requires sound rock footings to ensure it is stable.

Concrete arch dam: The arch dam is the most elegant of all the dam types and relies on the principle of the arch, a parabolic structure, which redirects all forces applied to it along the line of the parabola. In consequence all these forces are resolved into compressive stress and there are no tensile stresses. The forces it supports are, by this means, transmitted to the base of the arch (see Fig. 3.4). In the case of an arch dam this will be the rock faces of a valley wall. The earliest known arch dams were built by the Romans. An arch dam can only be built in a steeply sided rock ravine where both the sides of the ravine and the underlying substrata are sound rock formations. The arch is built on its back, bowing upstream, with the sides bedded into the rock faces

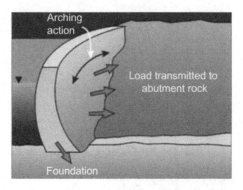

Figure 3.4 Schematic of an arch dam. Source: US Department of Homeland Security.[3]

of the side of the ravine. The arch then relies on the rock of the valley walls to support the weight of water pressing down on it. The concrete arch dam is the most precisely engineered of all dam types and requires less material than other types and where it can be built it is the strongest of dams. They are also among the tallest of dams. Most of the world's dams over 250 m high are concrete arch dams. In addition to the single arch dam, there are also multiple arch dams that rely on buttresses to support the forces on the arches in intermediate positions. In the United States, arch dams are classified according to their height with low dams less than 30 m tall, medium dams 30–91 m tall and high dams more than 90 m tall.

Steel dams: Steel dams use a steel framework to provide the structure of the dam, with sheets of steel applied to the upstream face to make a watertight barrier. The structure is usually similar to that of a buttress dam with sloping struts providing the support on the downstream side. These dams are relatively simple to build but the cost of steel makes them expensive and they are extremely rare today, though some were built at the beginning of the 20th century. They are costly to maintain too.

Timber dams: Timber dams are simple structures constructed from wood. Using wood makes construction relatively simple but they can only be used for low head, small hydro-projects. This type of dam was popular in the United States during the 19th century but is rarely found today except in relatively low technology applications.

Rubber dams: Rubber dams are a modern approach to the construction of small dams and barrages. They are fabricated as long

---

[3]"Pocket safety guide to dams and impoundments".

inflatable cylinders that can be filled with either air or water and then placed in a waterway. (Some dams alternate water and air depending upon the season.) The rubber is impermeable while the flexibility of the material allows it to adapt easily to the contours of the river bed. Rubber dams can be long — the longest is nearly 200 m in length — but they are relatively shallow. There are thought to be over 2000 in operation around the world. However, they are only used for small-scale projects. Water level can be controlled by inflating or deflating the dam to raise and lower it. Small amounts of water can be allowed to overtop a rubber dam without damage.

CHAPTER 4

Hydropower Turbines

The hydropower turbine is the element of the power station that converts the energy contained in the water from a waterway into mechanical rotary power that can be used to drive a generator and produce electricity. As such it is perhaps the most important element of a hydropower station. The hydraulic turbine is a simple, reliable and well-understood mechanical component, made from readily available materials. Most turbines are fabricated from iron or steel, or cast from bronze or other alloys. In the past wood was commonly used too. They are shaped to capture as much energy as possible from the water and this has led to a range of different designs, each suited to a different application. Some can extract the energy from high heads of water, others are better at exploiting low heads while yet others can take best advantage of variable rates of flow.

The history of the hydro-turbine is long. There may have been water-powered devices that were used to raise water for irrigation before the first millennium but documentary and physical evidence is scarce. The earliest known power devices, water wheels for grinding grain, were used by the Romans and were also known in China in the AD first century. Similar devices were common across Europe by the AD third century and could be found in Japan by the seventh century. Numbers grew at the start of the second millennium; the Doomsday Book of 1086 records 5000 in use in the south of England and there were large numbers recorded in France around this time. These early water wheels were made of wood and this continued to be the preferred material well into the second millennium. Iron was first used in the 18th century by an English engineer, John Smeaton, who studied and improved the efficiency of water wheels.

Early water wheels were often simple paddle wheels, the lower edge of which was placed into a flowing stream where the impulse of the flowing water against one side of the immersed paddle caused the wheel to turn. This is the basis for one group of turbines in use today,

Hydropower. DOI: https://doi.org/10.1016/B978-0-12-812906-7.00004-1

called impulse turbines, which exploit the kinetic energy contained in a jet of water. The development of these simple wheels led to a second method of extracting energy which involved adding a mass of water to each paddle at the top of the wheel and using its weight to rotate the wheel before releasing it again at the bottom of the wheel. These wheels required an artificial head of water to operate and may be considered as the early forerunners of the modern reaction turbine. Early water wheels were relatively inefficient but modern impulse and reaction turbines can be extremely efficient with the best converting 95% of the energy contained in the water into rotary motion. They can be stopped and started very quickly too, making them extremely flexible and they are capable of operating reliably for many years.

EARLY WATER WHEELS

The earliest water wheels, constructed from wood, existed in two basic types, the horizontal wheel and the vertical wheel. The horizontal wheel had a vertical shaft to which wooden paddles were attached to create a wheel in the horizontal plane. This orientation allowed the wheel to be coupled directly to a mill stone, which also traditionally rotates in the horizontal plane, without the need for gears or rotating joints. A jet of water was directed at the paddles on the wheel, forcing it to turn. Mills based on this type of water wheel, also known as a tub or Norse mill, were simple but relatively inefficient. The design was in use in both Asia and Europe at the beginning of the first millennium but its origins are unknown.

More common was the vertical wheel with a horizontal axis of rotation. The simplest version of this is the stream wheel, so called because it consists of a bladed paddle that is placed into the stream or river, as shown schematically in Fig. 4.1.

The stream wheel is the simplest vertical water wheel and the easiest water mill to construct. However its efficiency is very low at around 20%. The wheel relies entirely on the kinetic energy of the flowing water to drive the wheel and so works better with a faster flow. To this end they were sometimes mounted below bridges where the constriction caused by the bridge led to an increase in the rate of flow. Artificial means of constricting the flow were also used.

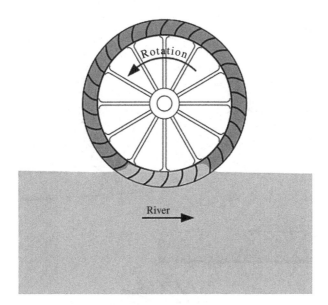

Figure 4.1 Schematic of a stream wheel. Source: Wikipedia.

Developing the concept further, some increase in efficiency was gained with an undershot wheel in which the water for the wheel is provided from a weir and strikes the wheel slightly higher up its circumference (an example of this can been seen in Fig. 1.2). More efficient still is the breastshot wheel (Fig. 4.2A) in which the water strikes the wheel midway up its height. This requires an even greater head of water but allows the wheel to capture energy both from kinetic energy of the flowing water and the potential energy of its mass through the pressure that applies to the wheel.

A further variation is the backshot wheel (Fig. 4.2B) in which the water is introduced to the wheel at or just before the summit or top. This type of wheel extracts (potential) energy entirely from the mass of water. The higher the head and the bigger the wheel, the more energy can be extracted. The way in which the water is introduced to the backshot wheel means that the wheel still rotates in the same direction as the undershot or stream wheel, and in the same direction as the water flowing into the tailrace from the bottom of the wheel. This aids its efficiency and makes it able to cope with flood conditions more easily. The overshot wheel (shown in Fig. 1.3) is a variant of the backshot wheel in which the water strikes the wheel just past the top. This causes the wheel to rotate in the opposite direction to that of a

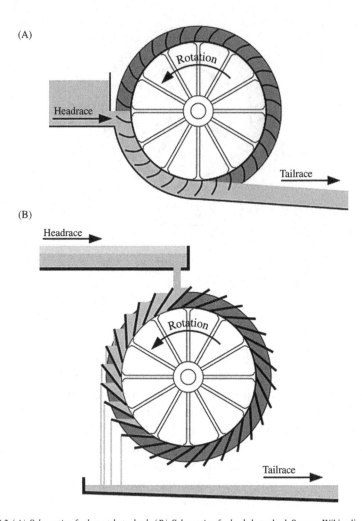

Figure 4.2 (A) Schematic of a breastshot wheel. (B) Schematic of a backshot wheel. Source: Wikipedia.

backshot wheel, and the opposite direction to the flow in the tailrace. Both types can achieve much higher efficiency than both the stream and undershot wheels.

TURBINES

Development of the water wheel during the 18th century allowed energy efficiency as high as 60% to be achieved. However the water wheel had limitations and during the 19th century even higher efficiency was achieved with the development of the turbine. The name turbine is derived from the Latin word turbo meaning 'spinning top' or 'that which swirls' and the name was introduced by the French engineer Charles Burdin. The key to these early turbines was that the device was enclosed and water was introduced tangentially, creating a swirling motion that improved energy efficiency. The most important of the early turbines was that designed by Benoît Fourneyron, a student of Burdin. His prototype was built in 1827 and he perfected it over the next decade. The design is still in use.

These new turbines had several advantages over water wheels aside from efficiency. They could operate with much higher heads of water and much higher water pressures and they turned at much higher speeds. This meant they could be smaller and more compact than water wheels.

Devices such as the Fourneyron turbine were reaction turbines which extracted energy from the pressure (or the mass) of the water. This led in turn to designs such as the Francis turbine, perhaps the most common in use today. Later in the 19th century, however, a new style of impulse turbine, one which extracted energy from the impulsive or kinetic energy of a jet of water, was developed. The most important of these new style impulse turbines is the Pelton turbine which is used to extract energy from very high heads of water at high efficiency. Pelton turbines are among the most efficient hydroturbines available.

The turbine superseded the water wheel for converting the energy of flowing water into mechanical energy and it is widely used today. However there are still applications for water wheels in some mini and micro hydropower installations, and they remain cost-effective for low technology developments if constructed from wood.

IMPULSE TURBINES

The impulse turbine is the descendant of the stream wheel. Like the latter, it uses the impulse or kinetic energy in a stream of water to drive the turbine and provide power. However while the stream wheel relies on the natural flow of water in a river or stream, an impulse turbine uses a powerful jet of water that is generated from a high head of water.

A tall column of water creates a high pressure at its base by virtue of the mass of the water pressing down on it and if this water under high pressure is released through a fine nozzle it will create a high-speed jet of water. If this jet is directed onto a bucket-shaped paddle on the circumference of a wheel, it will generate an impulse that will cause the wheel to turn. This is the basis for the impulse turbine.

The main type of impulse turbine in use today is the Pelton turbine, patented by the American engineer Lester Allen Pelton in 1889. However this was derived from an earlier design from 1866 by Samuel Knight who was inspired in his design by the high-pressure jets used for mining in California at that time. Both operate by firing a jet of water at buckets mounted onto a wheel. Both the Knight and the Pelton turbines operate at their highest efficiency, 95%, when the speed of movement of the bucket on the wheel rim is half that of the water in the jet directed into it.

A modern Pelton turbine has a set of spoon-shaped buckets mounted on the circumference of a wheel as shown in Fig. 4.3. The buckets are carefully formed so that the water from the jet entering the bucket changes direction and exits on the opposite side, transferring its momentum to the wheel as it does so. The water imparts most of its momentum to the wheel in the process and leaves the bucket at low speed. In many cases the Pelton turbine will have two rows of buckets mounted side by side around the circumference (with a gap between

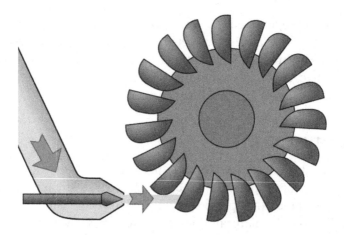

Figure 4.3 Schematic of a Pelton turbine. Source: Wikipedia.

them through which the spent water can escape), and these will be driven from two jets. This helps balance the lateral forces on the wheel and maintain smooth operation.

The Pelton turbine is generally used where a high head of water is available and the flow rate is relatively low. For large hydropower plants, they are normally used when a head height of more than 450 m is available but they can be used for heads a low as 200 m. The maximum head for a single Pelton turbine is generally 1000 m. Beyond that the fall will normally be divided into two sections with one turbine extracting energy from each drop.

A simple Pelton turbine will have one jet or nozzle, though as noted above two are often preferred, but power output can be increased by using up to four nozzles directed at the same wheel. Most Pelton turbines are mounted vertically with a horizontal axis of rotation, but they can be mounted with a vertical axis so that the turbine lies horizontally. A key feature of all Pelton turbines is that they must operate in free air, with no part of the wheel submerged. If part was submerged, it would cause drag on the wheel and waste energy. The turbine must therefore always be positioned about the water level at the bottom of the head of water being utilised for power generation. This reduces the available head slightly, but the difference will be minimal for a high-head plant.

The speed at which a Pelton turbine rotates will be determined by both the flow rate of water directed into its buckets and the load into which it is feeding. If load falls, the turbine will speed up. This can be controlled by reducing flow through the nozzles. The optimum efficiency is achieved when the turbine is operating at between 60% and 80% of maximum load. The largest Pelton turbines can provide a generating capacity of up to 400 MW while the smallest are only a few centimetres in diameter and generate 5 kW or less.

A second type of impulse turbine available for high-head use is the Turgo turbine. This is similar to the Pelton in design, but whereas with the Pelton turbine the water jets are in the same plane as the turbine wheel, in the Turgo turbine the jet strikes each bucket from one side and then exits the turbine at the other side. The Turgo turbine can handle higher flow rates than the Pelton turbine but is generally more difficult to construct. It is normally used for medium-head applications between those best suited to the Pelton and those more suited to reaction turbines.

REACTION TURBINES

Whereas the impulse turbine exploits the kinetic energy and momentum of flowing water, a reaction turbine relies on a combination of the kinetic energy and the pressure of the water generated from a head to drive it. This can be considered as the successor to an overshot or backshot wheel, but modern reaction turbines bear no resemblance to their predecessors.

A typical modern reaction turbine will be enclosed within a casing and the water is introduced around the sides of the casing, entering tangentially from all sides. The casing feeding the water into the turbine forms a spiral to create a swirl that helps animate the turbine. As the water enters the turbine, the pressure will act on the blades, forcing them to turn and the shape of the blade then forces the water to exit axially. The water is carried away through a channel called the draft tube which is designed to add suction to the forces acting on the turbine rotor, helping to transfer more energy to the machine.

There are several different types of reaction turbine. The most popular, accounting for 80% of all hydraulic turbines in operation, is the Francis turbine. This can be used in almost every situation but for very low heads propeller turbines and Kaplan turbines are frequently preferred. For all heads up to 450 m a reaction turbine is often the preferred design although there is some overlap of reaction and impulse turbines between 200 and 600 m.

FRANCIS TURBINE

The Francis turbine was developed by an American engineer James Bichens Francis around 1855. His design is extremely flexible and can be tailored to different head heights and flow rates. In operation the turbine must be completely immersed.

One of the key characteristics of the Francis turbine is the fact that water changes direction as it passes through the turbine. The flow enters the turbine in a radial direction, flowing towards its axis, but after striking and interacting with the turbine blades it exits along the direction of that axis. It is for this reason that the Francis turbine is sometimes called a mixed-flow turbine. In order for it to operate efficiently, water must reach all blades equally and flow is controlled by a

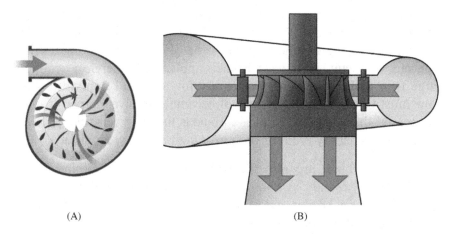

(A)   (B)

Figure 4.4 (A) Top view of a Francis turbine. Source: Wikipedia; *(B) Side view of a Francis turbine.* Source: Wikipedia.

casing which curls around the turbine in a spiral shape. This casing is called the volute (or sometimes simply the spiral) casing. The casing feed water through a set of valves and fixed blades into the moving blades of the turbine rotor. Top and side schematics of a Francis turbine are shown in Fig. 4.4A and B.

The blades of a Francis turbine rotor are carefully shaped to extract the maximum amount of energy from the water flowing through it. Water should flow smoothly through the turbine for best efficiency. The force exerted by the water on the blades causes the turbine to spin and the rotation is converted into electricity by a generator. Blade shape is determined by the height of the water head available and the flow volume. In general each turbine is designed for the specific set of conditions experienced at a particular site.

When well designed, a Francis turbine can capture 90%–95% of the energy in the water. While much of the energy capture is through reaction to the pressure of the water, there is also an important element of impulse transferred to the turbine blades too in consequence of the kinetic energy of motion of the water. This motion is generated by the spiral casing and the gates that feed the water into the turbine. The ratio of the two for a well-designed Francis turbine is probably around 1:1.

The Francis design has been used with head heights of from 3 to 600 m but it delivers its best performance between 100 and 300 m.

Flow rate is often the limiting factor for a given head. As the head height rises it increases the pressure at the base of the water column, and the size of the turbine must fall for a given flow, making fabrication more difficult. High-head Francis applications therefore require a large flow to be successful. Conversely for low-head applications the flow must be low or the turbine will become excessively large. It is for this reason that while the Francis turbine is the most versatile, different designs are generally used for both very high and very low heads.

Francis turbines are also the heavyweights of the turbine world. The largest, found at both the Itaipu power plant on the Brazil–Paraguay border and at the Three Gorges Dam in China, have generating capacities of 700 MW each.

PROPELLER AND KAPLAN TURBINES

If the head of water becomes too low, the rotational speed of a Francis turbine falls and with it efficiency. For low-head applications an alternative turbine type is required and the most successful is the propeller turbine. A propeller turbine looks much like the screw of a ship but its mode of operation is the reverse of the ship's propulsion unit. In a ship a motor turns the propeller which pushes against the water, forcing the ship to move. In the hydropower plant, by contrast, moving water drives the propeller turbine to generate power. A cutaway of a typical propeller turbine generator is shown in Fig. 4.5.

Propeller turbines are most useful for low-head applications such as slow running and lowland rivers. The head of water is typically 10 m or less. Their efficiency drops off rapidly when the water flow drops below 75% of the design rating, so plant designers often use multiple propeller turbines in parallel, shutting down some when the water flow drops in order to keep the remaining turbines operating at their optimum efficiency.

A variant of the propeller turbine is the bulb turbine which can be used under extremely low-head conditions (Fig. 4.6A). In this design the turbine is integrated with a water-tight generator and enclosed in a bulb-shaped container through which the water passes. The turbine rotor may have fixed blades but sometimes they are adjustable. Water flows into one end of the bulb-shaped container – called the nacelle – and out of the other, with no change of direction. The use

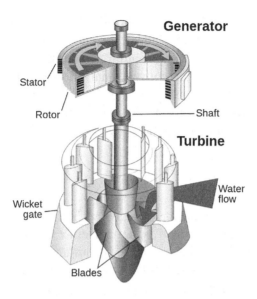

Figure 4.5 Cutaway of a propeller turbine generator. Source: US Corps of Engineers.

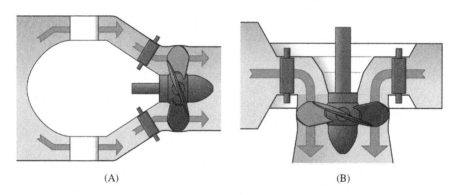

Figure 4.6 (A) Schematic of a bulb turbine. Source: Wikipedia; *(B) Schematic of a Kaplan turbine.* Source: Wikipedia.

of the nacelle helps concentrate the flow to maximise energy capture. The bulb turbine has commonly been used in tidal power plants.

There are cases where even though the flow varies in a river, multiple turbines cannot easily be used. In these, an alternative called the Kaplan turbine has sometimes been utilised instead. This low-head variant of the propeller turbine was developed by Viktor Kaplan in

1915. A schematic of a Kaplan turbine is shown in Fig. 4.6B. Its primary feature is a set of blades that can be adjusted to maximise efficiency under different flow conditions. These turbines can operate at higher flow levels than conventional propeller turbines and are suited to heads of between 10 and 50 m. A variation of the Kaplan turbine is the diagonal flow turbine which can operate at higher rotational speeds than the Kaplan and is suited to higher heads.

DERIAZ TURBINE

An additional reaction turbine available to hydropower plant designers is the Deriaz turbine invented by Paul Deriaz. This has blade shapes similar to a Francis turbine but these blades are adjustable and so can be adapted to different flow rates. The Deriaz turbine is another mixed flow turbine in which water enters from the size but exits along the turbine axis. It is best suited to head heights of between 20 and 100 m, in between the ranges of Francis and Kaplan turbines.

CHAPTER 5

Hydropower Generators

The generator is the part of a power plant that converts rotary motion provided by a device such as a hydro-turbine into electrical energy. The machine exploits a phenomenon discovered during the 19th century by the English scientist Michael Faraday; when a conductor moves through a magnetic field a current is generated in the conductor. The design of generators based on this principle has been refined over a period of more than 140 years and modern generators are extremely sophisticated and efficient electromechanical devices. Mechanical to electrical conversion efficiency for high-performance generators is only one or two percentage points short of 100%.

The earliest devices of this type were called dynamo-electric machines, named by Werner Siemens, the inventor of the first generator, in 1867. This name was soon shortened to dynamo. A dynamo is a machine that generates a direct current (DC) from mechanical rotation. Later, machines that could provide an alternating current (AC) were designed because it proved easier to distribute over large distances. These AC devices were originally called alternators but today they are usually called generators.

Early dynamos and alternators had coils that rotated in a stationary magnetic field generated by a permanent magnet. While relatively simple to build, this design proved to be a less efficient arrangement for larger machines and the generators that are used in large power plants including hydropower plants today reverse this arrangement; the magnet rotates and generates a current in stationary coils. Using a permanent magnet as the rotating element is often cumbersome. Instead, the rotating part of the machine, called the rotor, is an electromagnetic coil that generates a magnetic field when it is 'excited' by passing a DC electric current through it. This magnetic field induces a current in the stationary coils, called the stator, of the generator. The excitation current for the rotor may be provided externally from a second, small generator, or it may be provided by a secondary generator mounted

Hydropower. DOI: https://doi.org/10.1016/B978-0-12-812906-7.00005-3

onto the same shaft as the main rotor. This is known as self-excitation. The excitation current can be varied in order to control the power generated in the stator.

The excitation field provided by the rotor draws power that is dissipated in the stator. Some of this power emerges as heat within the stator. Losses from this source are typically only around 1% of the power of a large generator but even so this can amount to several megawatts of heat energy. This heat must be removed in order to prevent the coils overheating. There are other sources of energy loss too but total losses usually amount to less than 2% of total power output.

With such large quantities of heat being generated, cooling systems form a vital part of a modern generator. Cooling of large generators can be carried out using three principle coolants, air, water and hydrogen. Water is the most effective but also the most costly to implement, particularly in the rotor of a generator. Hydrogen is a very efficient coolant and is often used to cool the rotors of large generators but it is not normally found in hydropower generators. Instead, these use either air or water cooling. Small generators for hydropower plants will usually be air cooled, with fans used to circulate the air. For larger machines, the rotor is air-cooled and the stator water-cooled.

GENERATOR POLES AND ROTATIONAL SPEED

When a simple electrical coil rotates in a magnetic field produced by a single permanent magnet with two poles (north and south), the current that is produced in the coil is an alternating current that varies through one full 360-degree cycle for each complete rotation of the coil. This same relationship holds for a generator in which the rotor is an electromagnet that produces two magnetic poles. In most countries around the world, the standard frequency for the grid AC supply is either 50 Hz or 60 Hz. If the supply is to remain stable, large generators must supply power to the grid at exactly this frequency. In order to generate power at one of these frequencies with a two pole rotor, the rotor of the generator must turn at either 3000 rpm (50 Hz) or 3600 rpm (60 Hz).

This is a relatively high rate of rotation and not all turbine generator systems will be able to operate efficiently at such high speeds. In particular many types of hydro-turbine will turn at a much lower speed in order to extract energy from the water as efficiently as possible. It is feasible to increase the rotational speed by coupling the turbine to the generator through a set of gears. However the use of gears results in a loss of efficiency as well as being a weak point in the drive train of a generating system and while they might be used for small machines, most large machines are directly coupled to their generators. This means that the generator itself must be able to operate at lower speed.

In order to reduce the speed of rotation of a directly coupled generator there is an alternative solution, to increase the number of poles of the generator rotor. If the number of poles is doubled to four then there will be two complete AC cycles induced in the stator of the generator for each rotation of the rotor and therefore the rotor need only rotate at half the speed, 1500 rpm or 1800 rpm, to synchronise with the grid. The number of poles can be increased further in order to match slower rotational speeds. For example a 50-Hz generator coupled to a turbine rotating at 166.4 rpm will have 36 poles. Slower turbines will require generators with more poles still.

A feature of these low-speed hydropower generators is the use of salient poles in the construction of the rotor. This design, which is shown schematically in Fig. 5.1, has projecting poles made of laminated steel around which the coils that form the poles are wound. This arrangement is more convenient for generators that have multiple poles. For large, high-speed generators, the rotor will use a different, non-salient pole construction which is more robust when only two or four poles are required.

While the generator is designed to match the speed of the turbine, the speed of the combined system must still be controlled accurately if the system is grid connected. Most hydropower plants are connected as

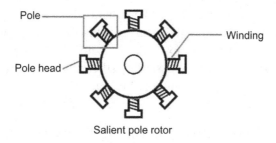

Salient pole rotor

Figure 5.1 Schematic of a salient pole rotor. Source: Wikipedia.

synchronous generators which means that they supply AC power at exactly (or very close to) the grid frequency. In order to maintain the speed accurately, the rotation of the turbine is controlled by a device called a governor. The governor, in turn, controls the degree of opening and closing of gates through which water enters the turbine. Opening the gates further allows more water through and the turbine will speed up. Increasing the closure of the gates will restrict the water flow further and the turbine will slow down.

The governor is essentially a feedback system that monitors the rotational speed of the shaft and then applies a correction if the speed is too high or too low. Early mechanical water wheel governors were in use by the 1860s and these slowly evolved as technology changed. Modern governors are computer-controlled speed control systems but they provide essentially the same service as early mechanical governors.

GENERATOR ORIENTATION

Another important consideration in generator design is the orientation of the generator. While most power plants utilise generators that are mounted with their rotational axes horizontal, in many instances the shaft of a hydropower turbine will be vertical, with the generator sitting directly above the turbine. This results in the weight of all the rotating elements, including both the generator and the turbine, being supported through a single bearing. This type of bearing must support the mass vertically while being able to rotate in the horizontal plane. One of more guide bearings above or below the support bearing will keep the shaft in position.

When the shaft is vertical, there are three possible positions for the thrust bearing, at the top, above the generator, in which case the complete assembly hangs from the bearing (this is often called a hanging bearing) at an intermediate position just below the generator, or lower still, at the top of the turbine assembly (these are often called umbrella bearings). Wherever it is situated, the load supported by the bearing must be transmitted to the foundations of the powerhouse.

High-head Pelton turbines usually have a vertical turbine wheel with a horizontal axis. However many Francis turbines have a vertical shaft. In addition low-head turbines such as propeller and Kaplan turbines will frequently be oriented with the shaft vertical.

GENERATOR TYPES

Most generators used in hydropower plants are a type known as synchronous generators. As discussed above, these are designed to be synchronised with the grid and they produce power at exactly the grid frequency. Synchronous generators, particularly large ones associated with large hydropower plants, provide an important service to the grid in addition to feeding it with electricity. Their ability to provide power at the grid frequency helps to maintain stability on the grid in the face of fluctuations in both demand and supply from other sources. Large hydropower turbine generators, which are physically massive, can also provide the grid with a form or inertia by virtue of the momentum associated with these large spinning masses. The inertia also helps stabilise the grid in the event of momentary failures or interruptions. In addition, a large angular momentum helps stabilise the turbine as the water flow rate varies and this momentum helps smooth out abrupt frequency changes.

There are alternative types of generator that can be used for hydropower generation. Some small hydropower plants use devices called asynchronous generators. These are very simple generators, sometimes motors that are run as generators and they cannot provide synchronised power. Instead they need the grid to provide their frequency stability and synchronisation. These generators tend to reduce the frequency stability of the grid rather than increasing it because they add an inductive load.

Another type of generator that is increasingly being used in small- and medium-sized hydropower developments is the variable speed generator. This does not seek to turn at the grid frequency but instead will rotate at a speed commensurate with the highest efficiency; when the flow rate increases, the rotational speed will increase and when the flow rate decreases, then so will the speed of the generator. This type of generator does not feed power directly into the grid. Instead it will use a solid-state power converter that converts the variable frequency AC into DC power and then converts this DC power back into AC at the grid frequency. Since this type of generator relies on solid-state

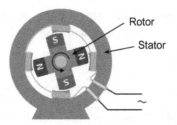

Figure 5.2 Schematic of a four-pole permanent magnet generator. Source: Wikipedia.

power devices, its power handling is limited to the levels that these devices can handle, which today is normally in the tens of megawatts for generators. However this type of generator is capable of providing grid support in the same way as a conventional synchronous generator.

While large multipole generators use conventional salient pole rotors, some modern variable speed generators use multipole permanent magnet generators. A schematic of a simple four-pole permanent magnet generator is shown in Fig. 5.2. This type of generator is built using rare earth magnets and is more efficient than a conventional generator. Permanent magnet generators are being used in offshore wind developments, where rotational speeds are often low and the speed varies with wind speed. Similar technology can be applied to small hydropower where the demands on the generator are similar.

CHAPTER 6

Small Hydropower

Small hydropower is a relatively imprecise description for a wide range of hydropower facilities. As was indicated in Chapter 2 the category includes any hydropower plant that has a capacity which is less than an upper limit that varies from country to country. The consensus upper limit is 10 MW but in the United States it is 30 MW and it can be higher in other countries too. The upper limit is important because it will affect whether a project can qualify as a new renewable generating plant. In general large hydropower plants are excluded from this category and do not qualify for grants that are available to support renewable projects. However some small hydro-schemes can qualify. The limit is normally defined by regulatory regimes and varies from country to country. In Sweden, for example, the upper size limit of a small hydropower plant that can attract support is 1.5 MW. In Italy the limit is 3 MW, in France 12 MW, in the United Kingdom 20 MW and in Canada 20−25 MW.

The global capacity of small hydropower was estimated to be 78,000 MW at the end of 2016 by the International Center on Small Hydro Power (ICSHP, part of the United Nations Industrial Development Organization). Another report estimated it to be around 110,000 MW.[1] Meanwhile the ICSHP has suggested that only 36% of the potential global small hydro-capacity has been exploited and around 139,000 MW remain. Other figures, such as those in Chapter put the potential much higher still with 2,949,000 MW of small hydro and 396,000 MW of micro hydropower available globally. Small hydropower generates 7% of the global renewable electricity according to the World Bank.

---

[1]Small Hydropower Market, by installed capacity − Global Industry Analysis, Size, Share, Growth Trends, and Forecast, 2015−2023, Transparency Market Research, 2015.

Hydropower. DOI: https://doi.org/10.1016/B978-0-12-812906-7.00006-5

Table 6.1 Small Hydropower Installed Capacity by Region, 2016	
Region	**Small Hydropower Capacity (MW)**
Asia	50,729
Americas	7863
Europe	38,943
Africa	580
Oceania	447
Source: *UNIDO*[2].	

Table 6.1 shows figures for installed small hydropower capacity for 2016 (the figures are for plants with capacities of less than 10 MW) broken down by region. Two regions stand out, Asia and Europe. Asia had the highest capacity 50,729 MW, while in Europe there was an estimated 38,943 MW. In the Americas the total was 7863 MW. The other two regions held tiny capacities by comparison. In Africa there were 580 MW while in Oceania the total was 447 MW.

Nationally, China dominates the small hydropower map with 51% of the total global installed capacity and this contributes heavily to the high installed capacity in Asia in Table 6.1. A further 16% is contributed by four nations, Italy, Japan, Norway and the United States.

SMALL HYDROPOWER BASICS

The design of a small hydropower plant depends very much on its size. The general designation 'small hydropower' still covers a wide range of plant capacities but it can be broken down further. Within this subdivision, *small hydropower* (note this is the same name as the overall designation) covers the range of 1–10 MW (or higher depending on the jurisdiction), *mini hydropower* includes plants between 100 kW and 1 MW while *micro hydropower* plants have capacities of between 1 and 100 kW. This latter range may sometimes be subdivided with into *micro hydropower* and *pico hydropower* plants. The latter then includes the range 1–10 kW.

The development of a plant within the *small hydropower* category (1–10 MW) will be approached in a similar way to that of a large

[2]World Small Hydropower Development Report, 2016, United Nations Industrial Development Organization.

hydropower project. While they are small compared to the largest hydropower plants, they are still relatively big generating stations and many of the same considerations and constraints apply. However size does affect economics. In this size range, dam construction is less likely to be cost effective so many will be run-of-river plants with a barrage to provide an intake structure for water to enter the hydro-system. Turbines may still be bespoke in the larger small hydro-schemes but at the smaller end of the range an off-the-shelf turbine is likely to be more cost effective. In most cases the turbine types used will be the same as those for large hydropower schemes.

The approach to the construction of a *mini* or *micro hydropower* plant of less than 1 MW in size usually involves a quite different design philosophy. At capacities below 1 MW, cost becomes the overriding concern. While traditional plant design may still be used, a range of novel techniques including inflatable barrages and at the lower cost end the use of cheap pumps as turbines may be employed to keep costs down. Site surveys and feasibility studies will be much more limited too.

One major difference between large and small hydropower is the breakdown of head height into categories. For a small hydropower plant a head above 100 m will be considered a high head and any project with a head of this or higher will employ a high-head turbine such as a Pelton turbine. For very small projects a Pelton turbine may be used at even lower head heights. Projects with heads of between 30 and 100 m are classified as medium-head scheme while anything under 30 m qualifies as a low-head plant.

Plant design will be much simpler in a small hydro-scheme. Most *mini* and *micro hydropower* plants will be run-of-river (or run-of-stream) and any intake structures, where used, are likely to be rudimentary to keep costs low. For larger plants a weir may be employed. Others will take water directly without any type of barrage. In many cases the turbine generator will be placed directly into the waterway.

If water is extracted from the river or stream it may be carried some distance through the equivalent of a headrace but more commonly it will be fed directly into a penstock-type conduit that carries it into the turbine. Penstock length can affect project costs significantly so this will be kept as short as possible.

Turbine types for small hydropower schemes will depend on head height; Pelton turbines for high-head, Turgo and Francis for medium-head and propeller and Kaplan turbines for low-head applications. Other turbines are also commonly used. These include a range of axial turbines, the cross-flow turbine which is a low-cost type of impulse turbine, the Archimedes Screw and the Gorlov turbine which is little like a vertical axis wind turbine that operates underwater. Simple paddle-type water wheels are also common. For very small applications cheap pumps can be used in reverse to make turbine generators. These are known as pumps-for-turbines or PATs and can be used with head heights of 13–75 m to build very cheap hydropower facilities. Small propeller turbines fitted with sealed generators can also be dropped directly into a stream for provide a hydro-generating system.

While larger small-hydro-schemes may use synchronous generators of the sort used in large plants, many small plants employ asynchronous generators that rely on the grid to help them control their speed of operation. In some cases these are simply motors being used in reverse to generate power. The efficiency of such small asynchronous generators is much lower than for large generators.

Small hydropower schemes tend to be relatively more expensive than larger schemes because costs of many of the components do not fall in line with size. In addition to the cost of a turbine and of any civil works, the cost of a grid connection can become a large component of the small hydro-scheme and a feasibility study can take a disproportionate share of the budget. This tends to push up to unit cost. The extra cost can still be economical if the small hydropower scheme is supplying power directly to consumers where it will be competing with the retail cost of electricity rather than the wholesale cost. Small hydropower can also be extremely effective in supplying power to remote communities far from a grid, especially when the alternative is diesel power.

SMALL, MINI AND MICRO HYDROPOWER PLANT DESIGN

The design and development of a *small hydropower* plant will follow the lines of a large hydro-scheme. The cost of small hydropower can be $5000/kW or more so a 1-MW plant is likely to cost perhaps $5 m. This is a significant investment for any developer. To ensure it is wisely spent, a feasibility study will be necessary and while this will be much more limited that would be the case for a major hydropower scheme, it will still need to provide details of the flow rate available at the site over a number of years, the flood conditions that are likely to be found at the site and some indication of the geological conditions where the power plant is to be built. A site survey will also indicate the best layout for the head of water available in the watercourse.

While many sites are completely undeveloped, it is common for some developments to be based at sites where there are existing structures, ancient water mills, for example, or an existing weir. This can make planning easier because there should already be a large amount of data available about the site.

The choice of turbine will depend on the site and head available. If an adequate head is available then a Pelton turbine might be suitable. More often a small Francis turbine is likely to be the most common choice. For very low heads a propeller or Kaplan turbine might be preferred.

Power plants in the *small hydropower* capacity range of 1−10 MW are likely to be grid connected, feeding power into the local distribution system. This will require an interconnection and the length of this will have an important influence on the overall plant cost. However it is possible for plants of this size to provide power to one or more isolated communities. This is more likely in countries of the developing world than in the developed. The type of generator may also be influenced by whether the plant is to be grid connected. A cheaper asynchronous generator might be used if the plant is to be connected to a strong, stable grid. However if the grid is weak, or if there is no local grid connection, then a synchronous generator may be preferable to ensure a stable supply.

When planning a *mini* or *micro hydropower* plant, design considerations start to change. While the larger plants in this category may be grid connected, many will not be. In addition, as noted above, cost becomes the key factor when design choices are being made. Site assessment is likely to be basic. For simple plants on small streams the flow available may be measured using a bucket while the estimate of head height will likely be carried out using very rudimentary methods.

Plants will almost always be run-of-river and construction materials used will be simple and readily available.

For plants in this size range head height is probably the crucial feature. The higher the head, the smaller the flow necessary for a given power output. If a small flow rate can be used, then a smaller, cheaper turbine can be installed and other structures such as the penstock can be smaller than for a site where a large flow must be exploited.

Fig. 6.1 shows a schematic of a *micro hydropower* scheme. The site is hilly and this will be typical of many schemes of this type because hilly and mountainous regions provide the best heads of water. The plant shares many of the characteristics of a large hydro-scheme. There is an intake structure from which water is taken through a canal that is similar in function to a headrace. This feeds water into a forebay where it is channelled into the penstock. The latter can be

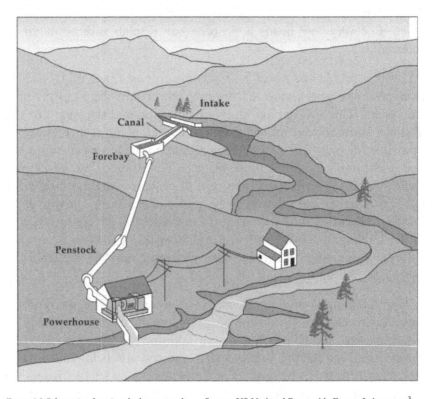

Figure 6.1 Schematic of a micro hydropower scheme. Source: US National Renewable Energy Laboratory.[3]

[3]Small Hydropower Systems DOE/GO-102001-1173 FS217 July 2001.

constructed from simple materials such as plastic piping. Water is carried through the penstock to a small powerhouse where it turns a turbine, generating electricity for local use. The water exiting the powerhouse is then returned to the river.

In this type of scheme the amount of power will be determined by the demand. This will usually mean that only a portion of the water from the watercourse is used for power generation. The remainder continues to pass down the stream or river. This is preferable environmentally because it means the local habitat is not disrupted.

Fig. 6.1 is typical of a *mini hydropower* or larger *micro hydropower* scheme. Smaller schemes are likely to be much more primitive. Some will use commercial turbines and others home constructed water wheels. Yet others will utilise devices that are essentially underwater windmills that can be placed in the flow of a stream or river and will provide an electrical output. For these smallest of plants, a head of more than 3 m might be considered a high head. The lowest head considered exploitable using the more conventional techniques is around half a metre. However for the submersible turbines a flowing stream that is deep enough to submerge it — perhaps 30 cm in depth — may be adequate.

TURBINES FOR SMALL HYDROPOWER PLANTS

The type of turbine that is chosen for a small hydropower plant will depend on the plant size and on the site conditions. For the larger *small hydropower* plants the choice will usually be between Pelton, Francis, Kaplan and propeller turbines. Typically for a *small hydropower* plant both Francis and Pelton turbines are available from a variety of manufacturers for generating capacities of up to 30 MW while Kaplan and a range of low-head axial turbines can handle capacities of up to 15 MW.

Depending upon the manufacturer, Francis turbines will be able to handle capacities from 500 kW to 30 MW and be able to operate effectively up to a head height of 300 m. While Francis turbines in larger plants are normally vertical, those in small hydropower plants can be horizontal too. This reduces the cost of the civil works necessary and simplifies the overall plant design. Meanwhile Pelton turbines for *small hydropower* plants have generating capacities of between 5 and 30 MW and will be able to handle head heights of between 60 and 800 m. These can be either vertically or horizontally oriented.

Specially designed Kaplan turbines can be used for low-head applications but many manufacturers also offer a range of axial turbines, often similar in concept to a bulb turbine and designed for easy installation. The key design feature for this type of turbine is that the water flows parallel to the turbine axis, allowing for a more convenient installation in a low-head plant. Many of these turbines will be directly coupled to generators but some can use a belt drive, which can again simplify construction. In other cases the generator is driven through a gearbox. Specific types of axial flow generator include the pit turbine in which the generator is placed in a pit behind the turbine (see Fig. 6.2A) and the S-type turbine in which the turbine sits at one end of an S-shaped draft tube through which the water flows (Fig. 6.2B).

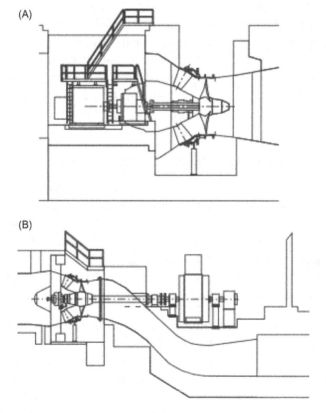

Figure 6.2 (A) Cross-section of a pit turbine, gearbox and generator. (B) Cross-section of an S-turbine and generator. Source: Voith Hydro.[4]

---

[4]Bult/Pit/S-Turbines and Generators brochure.

Another compact design of turbine generator is the Straflo in which the generator is built into the rim of the device in a shroud that surrounds the actual turbine runner. In this design the rotating turbine also acts as the generator rotor while the generator stator is built into the shroud. Straflo turbine generators for small low-head applications use permanent magnet generators which simplify the design.

Turbines for *mini hydropower* plants will often be similar to those of their slightly larger relatives. However the economics do not allow for bespoke turbine design so all will be off-the-shelf. In addition many will be sold as integrated turbine-generator packages, again helping to keep costs down. Simple installation and the need to keep civil works to a minimum will also affect both site choices and turbine design. Typical Francis turbines designed for *mini hydropower* applications will provide from 20 to 3000 kW of output at head heights of between 10 and 150 m. *Mini hydropower* Pelton turbines might have outputs of between 100 and 5000 kW and be designed for head heights of between 60 and 600 m. Similarly low-head axial flow turbines are available for outputs of 20–1500 kW and can exploit head heights of 2–26 m. Precise ranges and capabilities will vary from manufacturer.

New turbine types are also available for plants in this output range. The cross-flow turbine, also known as the Bánki-Michell turbine or the Ossberger turbine, is unique because the water does not flow either axially or radially through the turbine but runs transversely, across the turbine blades. This type of turbine is less efficient than more conventional designs but has a flatter efficiency curve as flow changes. Another option is the Gorlov turbine which is based on the Darrieus or egg-beater wind turbine. Meanwhile the Archimedes screw, based on the water pump design attributed to Archimedes, is a simple device that can exploit heads as low as 1.5 m.

When it comes to *micro hydropower* plants, yet more options are available. The largest in this range of 1–100 kW can use any of the turbines available for larger plants. Small Pelton turbines are common where there is a high head – in this case as low as 30 m – while for low-head designs, one of a range of compact, integrated, axial turbine-generator units can be used. There are also many novel turbines, some home-made, that can be used to provide power economically from a small stream or river.

One option is this range is to use a modern version of the ancient waterwheel. While this is not as efficient as some modern turbines it is simple to install and operate and the civil works needed for an installation of this type can be based on well-established principles that have been in use for centuries. Another is to use a pump. Water pumps are manufactured in large numbers and are very cheap. They can also operate in reverse so that if water is driven through the pump, then the motor will act like a generator and produce power. This is a cheap solution for many *micro hydropower* developers. Finally there are also a range of compact devices that work a little like underwater windmills. These devices have a propeller-like turbine that is fitted to a small generator in a sealed unit. The whole device is lowered into flowing water and produces electricity.

GENERATORS FOR SMALL HYDROPOWER

The type of generator used for a small hydropower plant will usually depend on the size of the development. The largest plants will overwhelmingly use conventional modern generator types, usually directly coupled to the turbine although some may drive through a gearbox so that a cheaper, higher speed generator can be used. There is also increasing use of multipole permanent magnet generators for small hydropower. While these are often more expensive than similar conventional rotor generators they are lighter, more compact and can be integrated into a small turbine-generator unit more easily. Several manufacturers now offer integrated turbine-generator packages that include a generator of this type.

As the size of the hydropower installation reduces, the cost of the generator becomes more of a consideration. The smallest plants will be designed around a cheap, readily available generator that operates at grid frequency. This may be driven through a gearbox or it may be belt driven – or both. Generators for these plants may be synchronous or asynchronous depending on the economics, upon whether the plant is grid connected and upon how strong the grid is in the region.

Another option to minimise costs is to use a motor as the generator. Cheap off-the-shelf motors can act as generators if driven in reverse and were widely used in the early days of the wind industry. They are asynchronous and need the grid to provide frequency stabilisation but can be used off grid but with less stable output.

CHAPTER 7

Tidal Power

The tidal rise and fall of seawater level along a coastline leads to the movement of large volumes of water in and out of coastal inlets and estuaries. This moving water can be used in the same way as the water flowing down a river as a means of generating electrical power and the technologies for the two share many features in common. Tidal movement is the result of the gravitational pull of the moon on the world's oceans and seas and it is extremely predictable. The movement varies from season to season.

The simplest way of exploiting the energy available from tidal motion is to build a tidal barrage across the mouth of an estuary or suitable inlet. The tidal changes in sea level will then cause water to flow cyclically backwards and forwards across this barrage. When the tide rises water flows from the sea into the estuary or inlet, passing through sluice gates in the barrage. At high tide the sluice gates are closed and when the tide ebbs the water behind the barrage is allowed to flow back to the sea through hydraulic turbines, generating power in the process.

The head height available for generation will vary with the state of the tide and a tidal plant will normally not start generating until some time after a high tide in order to obtain the optimum head for the site. In principle it is possible to generate when the tide is rising instead of when it is falling. Usually, however, ebb tide generation alone is preferred.

Exploitation of tidal motion has a long history and tidal mills with water wheels have been known for the best part of a millennium in Europe and elsewhere. The earliest record is from AD 900 but there will probably have been much earlier mills in operation. These tidal mills would impound water during the incoming tide, allowing the mill to operate for about 3 hours on each tide.

Hydropower. DOI: https://doi.org/10.1016/B978-0-12-812906-7.00007-7

Modern tidal power plants for electricity generation are relatively modern and also rare. Apart from some small plants built in China from the late 1950s onwards, the first commercial plant was built in France where it started operating in 1966. This remains one of the two largest operating tidal power plants and one of only a handful of commercial plants worldwide. The reason why there are so few plants is primarily down to the high cost of building a tidal barrage, the expense of which makes barrage tidal power plants appear uneconomical. There are also a limited number or sites where tidal plants of this type can be constructed, again limiting the potential. Nevertheless the tidal power plant remains of interest because of its long life and its reliability.

OPERATING TIDAL BARRAGE POWER PLANTS

Harnessing tidal motion to generate mechanical power has a long history. Tidal basins were being used in Europe to drive mills to grind grain before 1100. These plants were widely replaced when the Industrial Revolution introduced steam engines and fossil fuel but a few survived though there are none now operating commercially. The exploitation of tidal ebb and flow to generate electricity has been less well tried. The largest plant of this type (254 MW) is at Sihwa in South Korea, followed by La Rance on the northwest coast of France close to St Malo.

The 240-MW La Rance plant was built using specially devised bulb turbines. After La Rance, the third largest tidal barrage project is at Annapolis Royal on the Bay of Fundy in Canada. China has also developed some small-scale projects, of which the largest is at Jiangxia. Work on tidal power generation began in China in 1958 and there are thought to be seven projects in operation today with an aggregate capacity of 11 MW.

TIDAL POWER PLANT DESIGN

The main component of any tidal barrage power plant is the dam or barrage which is built across the mouth of a tidal estuary or inlet. This barrage is fitted with special sluice gates that are opened and closed during different stages of the tide. It is also fitted with hydropower turbines and these too are equipped with gates so that seawater can

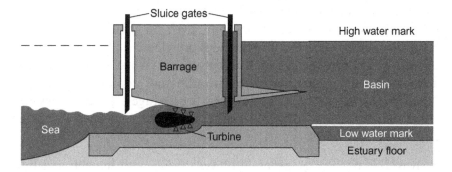

Figure 7.1 Cross section of a tidal barrage power plant.

either be allowed to pass through them or prevented from doing so as the tide changes. The cross-section of a typical tidal barrage power plant is shown in Fig. 7.1.

The simplest and most common form of power generation with a tidal barrage is ebb-flow generation. Under this scheme water is allowed to flow from the sea across the barrage and into the lagoon or basin behind it as the tide rises. Once high tide has been reached, the sluice gates within the barrage are shut, trapping the seawater within the lagoon.

The tide is now allowed to fall on the seaward side of the barrage to create a head of water across the barrage that will drive the trapped water through the hydro-turbines. The length of time during which the water is held will depend on the specific project but will normally be until the tide has fallen to around half its tidal range. At this point the gates closing the turbines are opened allowing the water from the lagoon to flow through them and back to the sea. Generation will normally continue until close to or after low tide.

When generation stops, the gates protecting the turbines are closed again and the sluice gates opened so that as the tide turns, water will once again pass the barrage into the lagoon. The cycle is then repeated through the next tide.

It is possible to reverse the mode of operation and generate power on the flood tide instead of the ebb tide. In this case the sluice gates are kept closed at low tide so that no water can pass. When the tide has risen by about half its range, the turbine gates are opened allowing water to flow through them and into the lagoon, generating power in the process. Generation continues until levels on either side of the barrage are similar when the turbine gates are closed and the sluice gates opened, permitting the lagoon to empty again. While this is operationally simply ebb generation in reverse, it is not commonly employed because it leaves the tidal basin behind the barrage exposed to low-tide conditions for extended periods, a situation that can have damaging environmental effects.

It is also possible to generate power during both the ebb and flow tide. The French plant at La Rance was designed to operate in this way but the plant in fact only operates in ebb generation mode. The main problem with two-way generation is that calculations suggest that economic gains are small and unlikely to be cost-effective because it necessitates the additional expense of either two-way turbines or two sets of turbines, one for each direction. On the other hand, it allows generation to take place for much more of the tidal cycle and leads to an overall lower peak power since the head of water that develops is never as high as with single direction operation. This would in principle allow smaller turbines to be used.

A further operational mode, one which has been employed at La Rance in France, is to use the turbines as pumps to pump additional water across the barrage. Pumping takes place at close to high tide, creating a larger head of water than would be available from the tidal range alone. It is possible to generate up to 10% more power using pumping than without it and the economics are attractive since the pumping takes place when the head height across the barrage is very small, so requiring little additional energy while energy is returned from a much higher head.

TWO BASIN PROJECTS

A conventional one-basin tidal barrage project can only generate power during a part of each tidal cycle. To get around this a variety of two basin projects have been proposed. This adds complexity but allows either continuous generation or generation for a longer period than a single basin design.

One type of two-basin design comprises two single tidal basins, each with its own barrage controlling the flow of sea water in and out.

These two basins are then connected by a channel into which turbines are fitted. In operation, one of the basins opens its sluice gates only close to low tide, keeping the water level within its lagoon as low as possible. Meanwhile the second opens its sluice gates towards high tide so that the water level within its lagoon is always high.

Water is then allowed to flow from one lagoon to the other through the channel linking the two. The flow rate and the capacity of the turbine within the channel is sized so that there is always more water in the high-water lagoon than in the low-water lagoon and that there will always be a head of water to drive the turbine.

The best developed project of this type was one proposed for construction near Derby in Western Australia. The project involved building barrages across two adjacent inlets and creating an artificial channel connecting the two basins formed by these barrages. A power station with turbines capable of generating 48 MW was to be stationed on this artificial channel. However the project was never built.

An alternative two-basin scheme design has a primary reservoir that acts like a normal ebb-flow tidal plant, generating power on the ebb tide. However on the seaward side of the primary basin is a second smaller basin. During the generation phase of the main basin, some power is used to pump water into the second basin, creating a storage basin from which power will always be available for generation whatever the state of the tide. The economics of such a scheme is relatively low at around 30% cycle energy efficiency.

TIDAL LAGOON POWER PLANT

Instead of building a barrage across an estuary, another approach is to create a man-made tidal lagoon. This involves building a sea wall to enclose an area of seabed. This can be located completely offshore or it can enclose an area adjacent to the coast, using the coastline as one side of the lagoon. Sluice gates are fitted into the sea wall and these allow water to enter the lagoon during the flow of the tide. At high tide the gates are shut and when the tide has partly fallen, the head of water left inside the lagoon is used to drive turbines installed in the lagoon wall to generate power.

The most significant plant of this type has been proposed for construction in Swansea Bay, off the coast of Wales. This U-shaped lagoon proposed for the project would have 16 turbines with a combined generating capacity of 320 MW. This is intended as a pilot scheme for a much larger tidal lagoon in the same part of the United Kingdom. The larger scheme would have a generating capacity of 3240 MW. Both projects are considered feasible but construction of the pilot scheme has yet to be approved suggesting the future of the scheme to remain uncertain.

TIDAL BARRAGE CONSTRUCTION TECHNIQUES

The construction of a tidal barrage represents the major cost of developing tidal power. As a result, much of the research work carried out into tidal power has focussed on the most efficient way of building the barrage.

Construction of the French tidal power plant at La Rance was carried out behind temporary cofferdams, enabling the concrete structure to be built under dry conditions. While La Rance was completed successfully using this approach, the method is generally considered too expensive as a means of constructing a tidal barrage today. There is also an environmental problem attached to completely sealing an estuary for the period of construction which might easily stretch into years. For that reason, such an approach is unlikely to be adopted for the future.

A novel approach suggested for the construction of a barrage across the River Mersey in England borrowed something from the construction of La Rance. The idea proposed here was to procure a pair of redundant bulk carriers, oil tankers for example, and sink them on the riverbed parallel to one another, sealing the ends and filling the enclosed space with sand to create an island. Concrete construction would be carried out on the island as if it were dry land. To create a watertight structure, diaphragm walls would be fabricated of reinforced concrete; the turbines and sluice gates required for the operation of the power station would subsequently be fitted to this concrete shell.

Once the first section of the barrage had been completed, the bulk carriers would be refloated, moved along to the next section and sunk again. This process would be repeated until the barrage had been completed. The Mersey barrage has not been built, so the efficacy of the method is yet to be tested.

Where an estuary is shallow, an embankment dam could be constructed instead of a concrete dam, using sand and rock as its main components. Sand alone would not make a stable embankment; wave erosion would soon destroy it. Hence, some form of rock reinforcement would be required on the seaward side. Concrete faces on both sides of the embankment could provide further protection. The sand needed for construction of such an embankment might be recovered from the estuary by dredging. Rocks could also be removed from the riverbed by blasting or brought to the site from elsewhere. Rock is a more expensive construction material than sand so its use would have to be minimised to keep costs as low as possible.

While all these methods have their attractions, the construction method most likely to be used to build a large barrage today would involve prefabricated units called caissons. Made from steel or concrete, the caissons would be built in a shipyard and then towed to the barrage site where they would be sunk and fixed into position with rock anchors and ballast. Some caissons would be designed to hold turbines, others would be designed as sluice gates and a third type would be blank. These would be placed between the other two types to complete the barrage.

TURBINES

The turbines in a tidal power station must operate under a variable, low head of water. The highest global tidal reach, in the Bay of Fundy in Canada, is 15.8 m and the mean tidal reach is probably half of this range; most plants would have to operate with much lower heads than this. Such low heads necessitate the use of a propeller turbine, the turbine type best suited for low-head operation. The fact that the head varies appreciably during the tidal cycle means that a fixed-blade turbine will not be operating under its most efficient conditions during the majority of the tidal flow; consequently a variable-blade Kaplan turbine is usually employed. As is the case with most low-head hydropower plants, tidal power plants usually employ a series of small turbines running along the barrage since these can exploit the available energy more effectively than a small number of large turbines.

The most compact and efficient design of propeller turbine for low-head applications is the bulb turbine in which the generator attached to the turbine shaft is housed in a watertight pod, or bulb, directly behind the turbine runner. The whole turbine generator assembly is then hung inside a chamber that channels the water flow through the turbine blades in order extract the maximum energy possible. The La Rance tidal plant employs 24 bulb turbines, each fitted with a Kaplan runner and a 10-MW generator. Bulb turbines were new when La Rance was built and construction of the plant involved some experimental work; of the 24 turbines, 12 had steel runners and 12 had aluminium bronze runners. Experience has led the operators to prefer the steel variety.

The turbines at La Rance were designed to pump water from the sea into the reservoir behind the barrage at high tide to increase efficiency. This was found to cause severe strain on parts of the generator and the design had to be modified. Work was carried out between 1975 and 1982. Since then the plant has operated smoothly and with high availability.

The more recent Sihwa tidal plant in South Korea also uses bulb turbines. In this case the plant is equipped with ten 26-MW bulb turbines with variable-blade propeller units. This power plant is built into a sea wall erected in 1994 to create an inland lagoon where water was collected for irrigation. Since then industrial pollution of the lagoon has made the water unusable. The tidal plant forms part of a scheme intended to flush the lagoon to reduce pollution levels. Unlike most other tidal barrage plants, it generates on the flow tide.

An alternative to the bulb turbine is a design called the Straflo turbine. This is unique in that the generator is built into the rim of the turbine runner, allowing the unit to operate in low-head conditions while keeping most of the generator components out of the water. A single large Straflo turbine generator was installed at the Annapolis tidal power plant at Annapolis Royal in the Bay of Fundy, Canada. This 18-MW unit is the only one of similar size that has been built, so experience with the design is limited.

The speed of a conventional turbine generator has to be closely regulated so that it is synchronised with the electrical transmission system to which it is attached. In order to aid frequency regulation under the variable conditions of a tidal power plant, a set of fixed blades called a regulator are often placed in front of the turbine blades to impart a rotary motion to the water. The use of these blades in conjunction with a variable-blade Kaplan turbine provides a considerable measure of control over the runner speed.

In small applications where such tight speed control may not be essential and where costs are critical, it may be possible to use one method of control — either a variable-blade turbine or a regulator — rather than both. An isolated unit that does not connect into the grid could operate without regulation.

An alternative option is to use a variable speed generator. This electronic solution will permit the turbine to run at its optimum speed under all conditions while delivering power at the correct frequency to the grid. This allows some efficiency gains. However the solution is more costly than a conventional generator with mechanical speed control of the turbine. Variable speed generators are being used on some hydropower schemes today. Capacity is limited but that is unlikely to be a problem with a tidal power plant where the unit size is generally small.

CHAPTER 8

Pumped Storage Hydropower

Pumped storage hydropower is an energy storage system based on the technology of hydropower. A conventional hydropower plant with a dam and reservoir involves an element of energy storage. The reservoir will store water during the rainy season and then deplete the reservoir during dryer periods. A pumped storage hydropower plant builds on this principle in order to be able to store and release electrical energy over a daily rather than a seasonal cycle.

In order to achieve this, the plant cannot rely entirely on the natural flow of water. Instead a typical pumped storage hydropower plant will have two reservoirs, one high and one low. When power is needed, water from the upper reservoir is released to the powerhouse of the plant when it drives the turbines and provides electricity. Meanwhile during periods of low demand spare electricity from the grid is used to pump water from the lower reservoir back into the high reservoir so that it is available again for generation.

Energy storage plants of this type rely on cheap electricity being available, surplus power that can be used to store water in the upper reservoir. Mountainous regions offer some of the best potential for pumped storage because of the high heads of water that can be exploited. One of the first plants of this type was built in Switzerland in 1882. The use of pumped storage grew during the early years of the 20th century but it was during the middle of the century that the major expansion took place, particularly in the United States, Europe and Japan. Many of the pumped storage hydropower plants that were built during this period were originally designed to complement nuclear power generation and large base-load coal-fired power stations. A conventional nuclear power plant operates most effectively if it generates power continuously, operating in what is known as a base-load power generation. Coal-fired power plants are also most efficient when they operate in base load too. However these plants are usually very large power plants with generating capacities in excess of 1000 MW and

Hydropower. DOI: https://doi.org/10.1016/B978-0-12-812906-7.00008-9

there may not be a need for their power during low demand periods such as the night-time. Pumped storage hydropower plants can use the surplus power from these large generating stations to store energy overnight by pumping water into the upper reservoir. This energy can then be made available to meet the peak demand the following day.

More recently there has been a growing need for electricity storage to help manage renewable electricity generation from intermittent technologies such as wind power and solar power. These are a good source of electricity while the wind blows or the sun shines but when wind or sun is not available, they can supply no power. The same principle that was applied to nuclear and coal power can be applied here too. As long as there is sufficient intermittent renewable capacity, surplus power available when the wind blows and the sun shines can be stored in a pumped storage hydropower plant. This energy can then be used to provide power when the renewable energy is not available.

Like hydropower itself, the technology for a pumped storage hydropower station is simple and well understood. The only unusual feature is that a pump is needed to move water into the upper reservoir in order to store energy. Many modern plants use reversible pump-turbines. While the technology presents no challenges, the civil works involved in the construction of a plant with two reservoirs, one high and one low, can be costly unless there are natural features that lend themselves to development. This is rare and more normally one of the reservoirs must be created. The construction work for such a plant makes the capital cost of the technology expensive. Once constructed, however, the lifetime of such a plant should be measured in centuries rather than decades.

The total global installed pumped storage hydropower capacity at the end of 2016 was 149,690 MW.[1] The largest number of plants is in Asia, where Japan and China both have significant capacity, and in Europe and North America. Pumped storage plants also account for around 97% of the total active electricity storage capacity available on grids across the world according to the US Department of Energy. Other technologies such as battery energy storage or flywheel energy storage devices are generally much smaller than pumped storage plants. Some pumped storage hydropower plants, provided they are

---

[1]International Hydropower Association Hydropower status report 2017.

built on a river, can also operate as conventional hydropower stations but many are designed specifically for energy storage alone. According to the Energy Storage Association, pumped storage plants in the United States can store around 2% of the nation's power output. In Europe the equivalent figure is 5% and in Japan storage plants can accommodate output from 10% of the country's power plants.

PUMPED STORAGE HYDROPOWER PLANT DESIGN

A typical pumped storage hydropower plant will have two reservoirs or lakes that are connected via tunnels and shafts through which water can be passed from one to the other. Within the tunnel system will be one or more hydro-turbines and pumps (these will usually be the same unit today) and there will be valves to control the flow of water from one reservoir to the other. A schematic of a pumped storage hydropower plant is shown in Fig. 8.1.

The simplest type of pumped storage hydropower scheme is based on a dam and reservoir power plant but with the addition of a lower reservoir at the end of the tailrace where water exits the powerhouse. Under normal conditions the plant will function simply as a conventional hydropower plant but with output regulated, so that most is supplied during periods of peak demand rather than operating as a base-load plant. In order to qualify as pumped storage, such a plant must also be equipped with pumps that can pump water from the

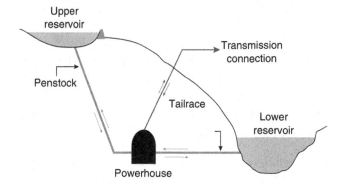

Figure 8.1 Schematic of a pumped storage hydropower plant. Source: Argonne National Laboratory.[2]

[2]USA Pumped Storage Hydropower: Benefits for Grid Reliability and Integration of Variable Renewable Energy.

lower reservoir back into the main reservoir when water levels are low and surplus power is available. Such plants are called on-stream integral pumped storage plants or pump-back pumped storage plants.

It is often convenient to make use of an existing dam and reservoir power plant when constructing a pumped storage hydropower plant because one of the reservoirs already exists. Many plants that are based on this design can operate both as pumped storage and generating plants. Frequently, however, both lakes need to be constructed. The largest pumped storage hydropower plant in the world, Bath County in the United States with a generating capacity of 3000 MW, was created by building two embankment dams, each damming an existing waterway to provide two new reservoirs. Water levels in both reservoirs vary during operation. The maximum head between the two is 400 m.

Sometimes it is more convenient to construct two man-made reservoirs that are not on any waterway (but some source of water will be needed to initially fill the reservoirs and to keep them replenished since there will be continuous losses). Old quarries have been used to provide one reservoir for such schemes. Plants of this type simply cycle water between upper and a lower reservoir and can only function as storage plants. They are called closed-cycle pumped storage hydropower plants.

One of the most difficult tasks when considering a pumped storage project is to find a location suitable for construction or exploitation of two reservoirs that are separated by a sufficient head to provide an efficient pumped storage plant. A high head is preferable for such a plant because this will provide more energy from a smaller flow than a similar low-head scheme. One option that has often been considered but rarely used is for the sea to provide the lower reservoir of a pumped storage scheme. For this to be effective there must be high cliffs adjacent to the shore and, crucially, either a lake or a site suitable for creating a lake at the top of the cliff. The largest plant of this type that exists today is a 30-MW scheme in Okinawa, Japan. Others have been proposed but none has yet been built.

Another alternative is to create a second reservoir underground. This is attractive because it has much less environmental impact than the creation of a new reservoir above ground but it is limited by the availability of a suitable underground reservoir. The most attractive sites are underground mine workings that have been abandoned. These exist in many parts of the world. However they have to be watertight to be of use; otherwise the water will leak away before it can be pumped back to the upper reservoir. Again, while such schemes have been proposed, none has been built.

Reservoir size is another factor that must be considered at the design stage, particularly if one or both are to be created. The storage capacity of the plant will depend on how much water the upper reservoir can hold. For a given set of turbines, the larger the reservoir, the more hours of power can be stored. Some plants are designed with up to 20 hours of storage capacity, others with as little as 4 hours. When operated on a daily cycle the round-trip efficiency of energy storage is 70%–80%.

Existing pumped storage plants have head heights of anything between 30 and 750 m. The minimum head height for a closed-cycle pumped storage plant is probably around 100 m but much higher heads are advantageous since the equipment needed and the volume of water that has to be cycled is lower. Plants with the highest heads often require either separate pumps and turbines or multistage combined pump-turbine units to be able to return water to the upper reservoir.

TURBINES AND MOTORS

A pumped storage hydropower plant must be able to both generate power from water running downhill and store energy by pumping water uphill. In the earliest pumped storage plants this was achieved by having separate pump-motor and turbine-generator units with each set mounted on their own shafts. Some very high-head plants that use Pelton turbines to generate power still use this configuration because the Pelton turbine while the most efficient for exploiting a high head, cannot act as a pump. For most, however, there was an element of redundancy with two motors and two turbines, one acting as a pump.[3]

By the middle of the 20th century there had been some design consolidation. Since a motor is simply a generator running backwards (and vice versa) the new configuration had a single shaft with a motor-generator at the top, a pump in the middle and a turbine at the bottom. The motor-generator could then function with both pump and

[3]The rotor in a pump is like a hydro-turbine but operating in reverse.

turbine. In many of these designs both the pump and the turbine was based on a Francis design. In spite of this, it was not until the middle of the century that single-unit pump-turbines began to be deployed in which the same rotor acted both as turbine and pump. These have now become standard for most pumped storage plants.

Pump-turbines based on the Francis turbine can be used for a range of head heights. However they can be less effective for low-head applications because in these the actual head height can change significantly as the upper reservoir is depleted and this will affect the generating efficiency of the turbine. A solution to this is to use a Deriaz pump-turbine. The Deriaz turbine is similar to a Francis turbine but has adjustable blades so that the blade angle can be altered as head height changes to maintain optimum efficiency.

Another advancement that is being used in some modern plants is the variable speed turbine generator. Pumped storage turbines have traditionally been synchronised to the grid, so each time they start up they must be synchronised before they can begin to supply power. A variable speed turbine generator, which has an electronic circuit to convert the output of the variable speed unit to that of the grid, has advantages both in terms of ease of synchronisation and efficiency because speed can be adapted to the flow rate to maintain optimum efficiency.

Whatever type of turbine is used, pumped storage hydropower plants are capable of reacting extremely quickly to a demand for power. For example the Dinorwig pumped storage hydropower plant in Wales has six 300 MW pump-turbines that can synchronise and reach full power in around 75 seconds. What is more, if the turbines are first spun up in air and synchronised ready to deliver power, the full capacity of the plant − 1800 MW − can be operational in 16 seconds.

In addition to being able to provide rapid response power to the grid, a pumped storage hydropower plant can help with grid stability in other ways. It can absorb surplus power rapidly as well as provide it and it can provide reactive load and voltage stabilisation. It can also supply what is known as spinning reserve, providing the grid with inertia that helps resist sudden changes in conditions.

CHAPTER 9

Hydropower Plants and the Environment

A hydropower plant is a renewable energy power plant that generates electricity using a renewable natural resource, water running down waterways as a result of rainfall. This makes it both sustainable and attractive as part of a strategy to combat global warming. In addition, hydropower and pumped storage hydropower can both be used to help balance other renewable energy sources such as wind power and solar power on the grid. However, like all power stations, hydropower plants have an environmental impact.

In the case of hydropower, particularly where a dam and reservoir are involved, the impact can be extensive. The impact can be negative but it is not necessarily so. In well-designed projects the benefits should easily outweigh the gains. Nevertheless, any project that involves the level of disruption that can result from construction of a major waterway scheme must be evaluated extremely carefully and all the implications taken into account before construction is authorised.

For a dam and reservoir project, the effects can be far reaching and so, too, are the considerations. What is going to be submerged when a reservoir is created? Will local communities be forced to move? What effect will the dam have on sedimentary flow in the river? How will downstream water flows be affected? What are the greenhouse gas implications? Large schemes have the potential to affect whole regions and require careful scrutiny at regional and sometimes at a national level. In some cases these effects will cross national boundaries too.

For a run-of-river scheme, the level of disruption is likely to be smaller but an extensive environmental study will still be required. Small hydropower schemes are rarely disruptive on the same scale as large hydropower projects and their impact is usually limited to the immediate locality so that decisions can often be made at a local rather than a national level. Even so, they should still be subject to the same criteria and standards.

Hydropower. DOI: https://doi.org/10.1016/B978-0-12-812906-7.00009-0

This problem is not new. Mankind has been altering waterways for at least two millennia and some of the early structures still exist. A rockfill structure dated from 1300 BC is in use in Syria and Roman dams can be found in Spain too. In the past, dams have been used to provide water for drinking and irrigation and to help control waterways. Only since the end of the 19th century has electricity generation been added to this list of uses.

While dams will always change the environment, in the past the changes wrought have generally been considered positive, providing an improvement in the living standards and conditions of the people affected. Greater environmental awareness coupled with some careless developments led to a change in perceptions towards the end of the 20th century and since then large hydropower schemes have been scrutinised much more carefully and critically. This prompted the World Commission on Dams (WCD) to look at what makes a good and what makes a bad hydropower project.

The work of the WCD resulted in 2000 in the publication of *Dams and Development: A New Framework for Decision Making*.[1] Since then a reappraisal of large hydropower has taken place resulting in greater acceptance of environmental considerations and many of the recommendations from the WCD have been broadly accepted. At the same time the link between water projects and the standard of living of people affected has been more widely recognised too. This is particularly significant in Africa where hydropower development is far behind other continents and standards of living are often very low too. Following on from the WCD report, in 2011 the International Hydropower Association published its *Hydropower Sustainability Assessment Protocol*, incorporating many of the WCD's ideas.

Both documents recognise that a hydropower project, particularly a large one, will be disruptive but also recognise that it need not necessarily be destructive. Environmental changes will take place, people may be displaced and habitats destroyed but all these effects *can* be handled sensitively so that, for example, displaced communities are given a stake in the project and offered much improved living

---

[1]*Dams and Development: A New Framework for Decision Making*, the World Commission on Dams, Earthscan, 2000.

conditions while inundated habitats are re-created alongside the area that is under water.

Even so, there remain critics, and there continue to be projects that do not follow the guidelines.[2] Getting the balance correct remains a challenge and while industry standards are ignored, large hydropower will continue to attract a level of disapprobation.

ENVIRONMENTAL ASSESSMENT

Today, in order to make a case for a major hydropower project, a thorough environmental assessment will usually be necessary and in most cases it will be mandatory. The effects of the project including any necessary resettlement, effects on biodiversity, the potential for seismic activity and the impact on areas downstream of the project will all have to be evaluated. Such a study should include proposals for the mitigation of any negative effects of the development. In many cases, particularly where international lending agencies are involved, a project will not be permitted to proceed unless the environmental assessment is favourable. This is equally true of public sector and private sector projects.

Unfortunately there are still projects that are not afforded the scrutiny they require. There are many reasons for this including narrow self-interest or political gain, corruption and unwillingness by national bodies to accept the judgement of international agencies. And, as the WCD itself acknowledged, there has been a tendency to play up the advantages and benefits of large hydropower schemes while playing down the damage they cause.

It is clear today that only a wide and impartial assessment can ensure that good projects are constructed while bad projects are rejected. This environmental assessment, where it is correctly carried out, will examine the potential impact of the construction and evaluate this against internationally agreed criteria. This will allow an objective decision to be made as to the value of the project and whether the benefits outweigh the cost. The most important environmental considerations that should be included in such an assessment are discussed in the following sections.

[2]Seven Sins of Dam Building, World Wildlife Fund, 2013.

RESETTLEMENT

The most divisive effect of any large hydropower project is likely to be the need to resettle people whose homes or communities will be destroyed. In building the Three Gorges Dam, the Chinese government moved 1.3 million people and more people might need to move if the reservoir banks become unstable. This represents one of the largest resettlement programmes ever undertaken for a hydropower project but even with much smaller numbers the result will be extremely disruptive for those involved.

Resettlement numbers can be large as in the case of the Three Gorges but how one can judge if they are too large? One way of at least comparing projects is to determine the number displaced for each megawatt of generating capacity installed. For the Three Gorges this ratio was 71. The Kedung Ombo dam in Indonesia, a 29-MW project that led to the displacement of 29,000 people had a ratio of 1000 people/MW. In contrast the Grand Coulee dam built in the 1930s in the United States had a ratio of 2.

It is, of course, vital to identify all those who are going to be affected. That may be straightforward when it comes to people whose homes are located in the region to be affected. But there may be many other people who forage in the area that will be inundated, for example, that are not so easy to identify. For a large project, there will be civil works besides the actual dam and reservoir that will also have an impact on local peoples. These must be included too.

If people are to be displaced, then a rule of thumb for modern developments is that they should be better off economically afterwards than before. More than that, people being moved should have a large say in where they are moved to and a stake in the project. If the project can provide wide community benefits, then people will support it. If not, then development should be questioned. Dealing with resettlement sensitively is also important. For example, in Laos the inhabitants of four villages were rehoused in one village but because of the relative size and influence of two of the villages, the new site suited these two villages but not the other two.

The situation becomes even more difficult when whole communities and their cultural and religious sites are likely to be destroyed. It is sometimes possible to move such sites; the most high profile example of this is the rebuilding of an Egyptian temple before inundation of the Nile behind the High Aswan dam in Egypt. However local cultural and religious beliefs may make such a solution unacceptable. International standards would require that such a project be rejected but in many cases, particularly in remote parts of the world, such considerations are all too easily ignored.

The cost of resettlement can be significant. This should be completely accounted for within the project but there is a danger that developers will attempt to minimise the resettlement costs in order to keep overall costs down. The result is likely to be that resettlement is not carried out in the sympathetic way it should.

BIODIVERSITY

Even if a dam and reservoir does not displace many people, it can still have an enormous impact on regional biodiversity if it affects a large area. The relative impact in this case can be crudely assessed by calculating the area inundated for each megawatt of generating capacity. This ratio for the Three Gorges Dam was 317 ha/MW while for the Grand Coulee dam it was 5 ha/MW. Meanwhile the Kompienga dam in Burkina Faso achieved the score of 1426 ha/MW, the highest of any recent project. Again this is only a broad indication of the effect since it will also depend on the type of terrain that is being submerged. However large shallow reservoirs will always have more impact by this measure than deep, narrow ones.

The greatest single danger to biodiversity is that a project will destroy the home of an endangered species. Since hydro-projects take a long time to develop, it is possible – provided that the danger is identified – to create a new habitat to replace the one that is threatened while work continues on dam construction. This can be relatively straightforward for plant species but can be much more difficult for animal species. However it is feasible. Indeed, some older projects are now introducing managed habitats that were not considered when each plant was initially built.

The effect of dam and reservoir construction on aquatic species is less obvious but can be equally dramatic. In France several rivers no longer support salmon as a consequence of dams and in China the Yangtze dolphin was declared extinct in 2006, partly as a consequence of hydropower developments along the river which prevented the animal's movement.

Other effects include the water in deep reservoirs becoming deoxygenated, affecting aquatic life that might otherwise live there. On the other hand the creation of a reservoir can provide new opportunities for fish species and large reservoirs can allow fish farms to develop, creating a new industry that did not previously exist.

There is a danger when assessing the impact on biodiversity of looking at individual species but ignore the cumulative effect that a project can have over a wide area and over a long time. Sometimes the effect is unforeseen and only manifests itself years after the project is complete.

GEOLOGICAL EFFECTS

There is a growing body of evidence that the construction of dams and inundation of reservoirs can generate seismic activity in the underlying strata as a result of the pressures generated at the surface. Such effects are normally only found with large dams, over 100 m high. The activity is generally short-lived but in some cases it can persist. An earthquake in Sichuan province in China in 2008 has been linked to a dam. This earthquake caused the loss of 80,000 lives. The pressures are not generated by the dam structure alone. The mass of water over the whole region of the reservoir will lead to significant subterranean pressure that did not exist before.

Another danger is of landslips in the region around the reservoir. A case of this type in Italy in 1983 caused a reservoir to overtop, leading to the loss of 2600 lives. Such landslips are not only potentially fatal, as in this case, but they also reduce the volume of the reservoir and therefore its utility. In addition they destroy additional habitat that was not initially affected by the project.

SEDIMENTATION AND DOWNSTREAM EFFECTS

All rivers carry a burden of small particles that are picked up by the flow and borne downstream with the water. When a river is dammed, this burden of sediment will often simply settle in the bottom of the reservoir and slowly fill it up. The reservoir for the Sanmen Gorge hydropower plant on the yellow river in China lost 40% of its volume to sediment in 4 years. While this is a dramatic case, most reservoirs have to deal with this problem on some scale.

Normally sediment deposition will eventually reach a steady state with enough sediment being carried past the dam to balance that being deposited every year. It is sometimes possible to wash sediment away periodically by opening the sluice gates of the dam. However sediment is made of abrasive particles and its passage through the plant's turbines will cause wear which may eventually lead to the need for repair or replacement.

The effects on the reservoir are important but many of the most important environmental effects of sediment deposition in a reservoir are felt downstream of the dam. One important consequence is likely to be the loss of sediment on downstream habitats which rely on it. When the Aswan dam was built on the river Nile it prevented vital sediment reaching the Nile delta. This sediment was the source of the delta's fertility and its loss led both to delta erosion and to a rapid increase in the use of artificial fertilisers. Today the habitat of the Nile delta is completely different to that known historically. Similar consequences can be found all over the world. For example, problems in the Black Sea with algae have been linked to loss of sedimentary material as a result of dams on rivers feeding into it, especially the Danube.

Another effect of lack of sediment in the water flowing past the dam is to increase erosion immediately below the dam site. The downstream erosion can be extremely damaging and the effect will be particularly noticeable immediately after a dam has been built. With time, the downstream terrain may be cleared of all material that can be readily eroded and the riverbed will end up being lined with obdurate rock. Elsewhere, rivers often narrow and deepen as a result of erosion. River banks can be undercut, damaging foundations of bridges and of other riverain structures. In the optimum case a new steady state is reached several years after dam construction but this is not always the case.

Against this, a dam on a major river such as the Danube or the Yangtze can help control flooding during periods of high rain. This can make downstream regions much safer and flood control is an important part of some dam projects. A dam can also make the river navigable upstream, where it was not before, which can be of benefit for local communities.

GREENHOUSE GASES

Hydropower projects are generally classified among the power generation schemes with the lowest greenhouse gas emissions. Typical greenhouse gas emissions are 10–13 kg/MWh, similar to that of wind power plants. However not all hydropower schemes are low emitters. Some can generate significant quantities of methane, a potent greenhouse gas.

Methane is produced when organic material collects in the bottom of a reservoir where the water is deoxygenated. Under these conditions anaerobic digestion takes place releasing methane gas. In order to prevent this, project developers are advised to try to remove as much organic materials as possible from the region to be inundated before submersion takes place, by felling trees and clearing undergrowth where possible. Even so it will be impossible to remove everything.

The Canadian utility Hydro Quebec, which has studied this effect, has found that methane production from reservoirs normally follows a predictable cycle. Production peaks between 3 and 5 years after the reservoir is filled. After 10 years emissions are no greater than for natural lakes. However there have been cases where much higher levels of methane emission have been detected.

The level of emissions may depend on the type of region in which a reservoir is created. Some recent research in the United States has suggested that in agricultural areas, farming effluent may be carried into reservoirs and so continue to feed the bacteria that produce methane.[3] The extent of such effects is largely unknown today.

INTERREGIONAL EFFECTS

Dam construction can lead to political disputes when rivers cross national boundaries. For example, the damming of the river Euphrates in Turkey has reduced water flow through Syria and Iraq and this has

[3]Methane Emissions May Swell from behind Dams, Bobby Magill, Scientific American, 29 October 2014.

led to frequent disputes. Reduced water flow when water is taken upstream for irrigation or drinking is one problem. Others may relate to sedimentation issues discussed above. In all cases, however, friction is likely unless great care is taken with such developments.

CHAPTER 10

The Cost of Electricity From Hydropower Plants

Hydropower plants are generally considered to be capital-intensive power projects because most of the cost is associated with the construction of the plant and very little with its operation. Large projects usually involve major civil engineering works with high labour and material costs and this represents the major part of the overall cost. In consequence of the high upfront cost, most of the funding must be available at the outset. Many large hydropower projects are multipurpose schemes that provide water for drinking and irrigation and new local facilities as well as electrical power. In the past the cost of building a hydropower plant of this type was often borne by the public sector and apportioning the costs between the different functions was not necessary. However since the liberalisation of electricity markets, which started towards the end of the 1980s, it has often fallen to private sector companies to fund them.

Some major schemes are still funded by the public sector and others are funded through financial support mechanisms such as those of the World Bank or through other international lending agencies. Increasingly, too, private investment is finding its way into hydropower. However financing is often complicated by the fact that at least some of the benefits of a new hydropower project, particularly in the developed world, will accrue to the government. These benefits include flood control and the supply of irrigation and drinking water. Projects that rely on private sources of investment can end up with their scope compromised in order to control costs, at the expense of those who might benefit from the local amenities such a project can provide.

Another factor that affects the economics of hydropower is the long life that can be expected from a well-designed project. While most power plants have useful lives of 30–40 years at most, a hydropower plant can continue to operate for over 100 years — or potentially longer — provided the turbines are maintained and periodically replaced. However financing of a project is unlikely to be possible over

Hydropower. DOI: https://doi.org/10.1016/B978-0-12-812906-7.00010-7

100 years, so costs will weigh heavily on the early years of the project. Once the cost of loans needed to build a hydropower plant have been paid, the cost of electricity from the plant is likely to be as cheap or cheaper than virtually all other sources.

The economics of hydropower are affected by the size of a power plant too. While the capital cost of a large plant is high, the relative capital cost of small hydropower plants tends to be larger than for larger plants. When examining the economics of a hydropower scheme, one of the most important parameters will be the cost of electricity from the plant. This will be a function of the capital cost, funding costs and operational and maintenance costs. It is common for an economic model called the levelized cost of electricity (LCOE) model to be used to arrive at a figure for the cost of electricity from power plants including hydropower plants. This allows the relative value of different types of generating plant to be compared.

LEVELIZED COST OF ELECTRICITY

The cost of electricity from a power plant of any type depends on a range of factors. First there is the cost of building the power station including extensive civil works in the case of hydropower and buying all the components needed for its construction. In addition, many power projects today are financed using loans so there will also be a cost associated with paying back the loan, with interest. Then there is the cost of operating and maintaining the plant over its lifetime. Finally the overall cost equation should include the cost of decommissioning the power station once it is removed from service.

It would be possible to add up all these cost elements to provide a total cost of building and running the power station over its lifetime, including the cost of decommissioning and then dividing this total by the total number of units of electricity that the power station actually produced over its lifetime. The result would be the real lifetime cost of electricity from the plant. Unfortunately such a calculation could only be completed once the power station was no longer in service. From a practical point of view, this would not be of much use. The point in time at which the cost of electricity calculation of this type is most needed is before the power station is built. This is when a decision is made to build a particular type of power plant, based normally on the technology that will offer the least cost electricity over its lifetime.

In order to get around this problem, economists have devised a model that provides an estimate of the lifetime cost of electricity before the station is built. Of course since the plant does not yet exist, the model requires that a large number of assumptions be made. In order to make this model as useful as possible all future costs are also converted to the equivalent cost today by using a parameter known as the discount rate. The discount rate is almost the same as the interest rate and relates to the way in which the value of one unit of currency falls (most usually, but it could rise) in the future. This allows, for example, the cost of replacement of a plant component 20 years into the future to be converted into an equivalent cost today. The discount rate can also be applied the cost of electricity from the power plant in 20 years' time.

The economic model is called the LCOE model. It contains a lot of assumptions and flaws but it is the most commonly used method available for estimating the cost of electricity from a new power plant. One particular problem is the level at which the discount rate is set. It is typical to use a discount rate of 5% and 10% in calculations. However in the middle of the second decade of the 21st century the actual interest rate is close to zero. For hydropower the length over which a loan will be financed can also have a major impact. Most power projects are financed over 20–30 years, similar to the expected plant lifetime. A hydro-plant can expect to operate for 50–100 years, maybe more. If a shorter financing period is used, it makes the overall cost of electricity higher during the period over which the loan is repaid. It then falls dramatically.

CAPITAL COST OF HYDROPOWER PLANTS

The breakdown of costs for a hydropower plant suggests that typically 60%–70% of the total is accounted for by the civil works. Equipment only accounts for 25%–35% while engineering and consultancy takes the remaining 5%–10%. Since the civil engineering portion of the project will involve considerable labour costs, overall costs will vary with these costs. Labour costs in some regions are likely to be much lower than in others.

Actual capital costs for hydropower plants vary widely but typical costs, based on published figures for recent plants, are between $1000/kW and $2000/kW. Many of these plants have been built in the developing world where labour costs tend to be lower than in the developed. The US Energy Information Administration (US EIA) has estimated that the cost of a new 500 MW hydropower plant in the United States, commissioned in 2016 and entering service in 2020 would be $2200/kW, just outside the upper limit of the range above.[1] The International Renewable Energy Agency (IRENA) in a report from 2012 put the capital cost of large hydropower projects at between $1050/kW and $7650/kW.[2]

Small hydropower plants tend to cost more than the larger projects because many of the costs do not scale with size. IRENA found the cost of small hydro-projects from a range of countries varied between $1300/kW and $8000/kW. However the actual costs of such projects will depend on both size and the type of technology being used. Very small projects are potentially cheap if they use simple technology such as pumps-as-turbines. However small projects can also be relatively expensive depending on the location and cost of technology. Meanwhile adding hydro-generating capacity to a dam that does not have turbines can be achieved for as little as $500/kW.

LCOE FROM HYDROPOWER PLANTS

Even with relatively high capital costs, hydropower can offer a low cost of electricity option. For example in the United States, the US EIA estimated that the cost of electricity from a new hydropower plant entering service in 2022 would be $63.9/MWh. Of the common fossil fuel−based technologies, this was only undercut by a natural gas combined cycle power plant without carbon capture and storage. Geothermal and onshore wind costs are also estimated to be lower.

This price will be based on some form of financing and loan repayment. However for plants in the United States that have paid off their loans, generation costs are estimated to be between $20/MWh and $40/MWh, undercutting virtually any alternative source.[3]

[1]*Assumptions to the Annual Energy Outlook, 2017*, US Energy Information Administration, 2017.
[2]Renewable Energy Technologies: Cost Analysis Series − Hydropower, IRENA, 2012.
[3]See the Pew Center web site, http://www.pewclimate.org/technology/factsheet/hydropower.

The cost of small hydropower (less than 50 MW) was estimated for a range of nations in a report from 2015 commissioned by the UK Department of International Development. This found that the levelized cost varied from around $50/MWh in India and Nepal to $100/MWh in African nations such as Zambia and Zimbabwe, up to $150/MWh or more in Palestine and Somalia.[4]

The report into the cost of hydropower published by the IRENA in 2012 found that the typical LCOE from large hydropower plants varied between $20/MWh and $190/MWh. For small hydropower, the range was $20/MWh–$100/MWh but for the very smallest of projects this could rise to $270/MWh.[5]

---

[4]Levelized Cost of Electricity, prepared by Bloomberg New Energy Finance for the UK Department for International Development, 2015
[5]Renewable Energy Technologies: Cost Analysis Series – Hydropower, IRENA, 2012.

INDEX

Note: Page numbers followed by "*f*" and "*t*" refer to figures and tables, respectively.

Printed in the United States
By Bookmasters